The Good and the Bad News About
QUALITY

QUALITY AND RELIABILITY

A Series Edited by

Edward G. Schilling

Center for Quality and Applied Statistics
Rochester Institute of Technology
Rochester, New York

Additional volumes in preparation

— The Good and the Bad News About

QUALITY—

Edward M. Schrock

Quality Control Consultant
Denver, Colorado

Henry L. Lefevre

Lefevre Enterprises, Inc.
Denver, Colorado

Sponsored by the American Society for Quality Control
Administrative Applications Division

Marcel Dekker, Inc.   New York and Basel
ASQC   Milwaukee

LIBRARY OF CONGRESS
Library of Congress Cataloging-in-Publication Data

Schrock, Edward M.
   The good and the bad news about quality / Edward M. Schrock, Henry
L. Lefevre ; sponsored by the American Society for Quality Control,
Administrative Applications Division.
        p.   cm. — (Quality and reliability ; 13)
   Includes index.
   ISBN 0-8247-7845-6
   1. Quality control. I. Lefevre, Henry L. II. American Society
for Quality Control. Administrative Applications Division.
III. Title. IV. Series.
TS156.S366 1988
658.5′62—dc 19   87-30514
                                                                CIP

MARCEL DEKKER, INC.
270 Madison Avenue, New York, New York 10016

Current printing (last digit):
10 9 8 7 6 5 4 3 2 1

PRINTED IN THE UNITED STATES OF AMERICA

About the Series

The genesis of modern methods of quality and reliability will be found in a simple memo dated May 16, 1924, in which Walter A. Shewhart proposed the control chart for the analysis of inspection data. This led to a broadening of the concept of inspection from emphasis on detection and correction of defective material to control of quality through analysis and prevention of quality problems. Subsequent concern for product performance in the hands of the user stimulated development of the systems and techniques of reliability. Emphasis on the consumer as the ultimate judge of quality serves as the catalyst to bring about the integration of the methodology of quality with that of reliability. Thus, the innovations that came out of the control chart spawned a philosophy of control of quality and reliability that has come to include not only the methodology of the statistical sciences and engineering, but also the use of appropriate management methods together with various motivational procedures in a concerted effort dedicated to quality improvement.

This series is intended to provide a vehicle to foster interaction of the elements of the modern approach to quality, including statistical applications, quality and reliability engineering, management, and motivational aspects. It is a forum in which the subject matter of these various areas can be brought together to allow for effective integration of appropriate techniques. This will promote the true benefit of each, which can be achieved only through their interaction.

In this sense, the whole of quality and reliability is greater than the sum of its parts, as each element augments the others.

The contributors to this series have been encouraged to discuss fundamental concepts as well as methodology, technology, and procedures at the leading edge of the discipline. Thus, new concepts are placed in proper perspective in these evolving disciplines. The series is intended for those in manufacturing, engineering, and marketing and management, as well as the consuming public, all of whom have an interest and stake in the improvement and maintenance of quality and reliability in the products and services that are the lifeblood of the economic system.

The modern approach to quality and reliability concerns excellence: excellence when the product is designed, excellence when the product is made, excellence as the product is used, and excellence throughout its lifetime. But excellence does not result without effort, and products and services of superior quality and reliability require an appropriate combination of statistical, engineering, management, and motivational effort. This effort can be directed for maximum benefit only in light of timely knowledge of approaches and methods that have been developed and are available in these areas of expertise. Within the volumes of this series, the reader will find the means to create, control, correct, and improve quality and reliability in ways that are cost effective, that enhance productivity, and that create a motivational atmosphere that is harmonious and constructive. It is dedicated to that end and the readers whose study of quality and reliability will lead to greater understanding of their product, their processes, their workplaces, and themselves.

Edward G. Schilling

Preface

The Good and the Bad News About Quality was conceived and initiated by Edward M. Schrock, a pioneer in the field of quality control. Ed, however, did not live to see it published. He was revising the book at the time of his death. I received the manuscript from Marion Schrock, Ed's widow.

Material in the book follows Ed's original work. I completed his revisions, introduced my own style, and clarified the technical parts. The basic contents, however, are the same despite two major rewrites and 2001 minor revisions.

Why me?

Ed was a consultant for the Coors Container Company at the time I managed their quality engineering department. During the five years we worked together, I developed an in-depth understanding of his ideas and philosophy. He also taught me how to find and corral maverick gremlins who delight in producing scrap, degrading quality, and initiating witch-hunts.

Most antigremlin tools involve statistics, but don't cringe. We approached the subject as managers, not as mathematicians. Executives need to know the language; statisticians need to know the formulas.

Technical material in this book was written for executives. If they understand it, anyone can. If they read it, everyone will. Management isn't dumb. They just don't read books with too many four-syllable words, confusing equations, and 12-line sentences.

The many examples scattered throughout this book come from observations made by Ed and me. We both survived decades of exposure to subjective decision making, fruitless witch-hunts, and professional wheel-spinning contests. We also observed masters at work—and learned from them.

We trust the reader will benefit from the disasters and successes we have described. It is better to learn from the errors of others than to make the same mistakes once.

Names of people and places have been changed to protect the innocent as well as the guilty. You, the reader, may know both.

Henry L. Lefevre

Contents

The Good and the Bad News About
QUALITY

1

The Good and the Bad News About Quality

Have you ever heard a dog purr or a cat bark? It can happen, you know.

Chromosomes and genes work almost perfectly with live household pets. This, however, is not necessarily true with toy animals whose barks and purrs come from mechanical inserts. If the insert for a dog looks the same as the one for a cat, someone is sure to mix them up. Then, a purring dog or a barking cat will probably reach the market place.

Quality conscious toy makers can design inserts that fit into their dogs but not their cats, and the other way around. Then, the toy animals will not confuse their owners with unorthodox sounds.

The good news is that we know how to economically manufacture high quality, reliable products. The bad news is that many manufacturing companies have not learned the good news.

Tragic events such as those associated with the Pinto rear-end crashes, the Bhopal chemical catastrophe, and the Three Mile Island nuclear accident are not inevitable. Steps can be taken to anticipate and initiate corrective action before disaster strikes.

Can companies avoid these quality problems?

Of course!

To do this, however, top management must don the armor of Saint George and slay the seven-headed dragon that gives birth to defects. The seven heads, each representing a management misconception, are listed below.

1. Quality control doesn't need high-level managers.
2. All defectives should be removed by inspection.
3. Reliable, high-quality products are uneconomical.
4. Quality is a departmental function.
5. Quality campaigns ensure quality.
6. Defects are due to lazy workers.
7. Good design and production ensure good products.

These misconceptions won't guarantee failure but they will make it probable.

DRAGON HEAD 1

Quality Control Doesn't Need High-Level Managers

Some companies rate their quality control manager between a janitor and the head of their typing pool. Does this imply that quality is Job #1? Would anyone shove an important assignment that far down the chain of command?

Most solutions to quality problems require help from many departments. Few fourth-level managers can ensure this type of team effort. Top rate, third-level managers (directors and junior vice presidents) have a fighting chance. Good, second-level managers (executive vice presidents) usually succeed. Company presidents get the job done.

DRAGON HEAD 2

All Defectives Should Be Removed by Inspection

How many of us can walk on water without knowing where the stones are? Can we expect quality control inspectors to do better?

In real life, an honest effort to find defects might catch 95 percent of them. If the defects are hard to detect, the number could be less than 90 percent. Inspection accuracy depends on the tools available, the inspector's attitude, the inspection procedures, and the quality characteristics being inspected.

Tools are normally controlled by budget, which in turn is controlled by top management. Never send an inspector dragon hunting with a rusty sword and a flammable suit of armor.

Attitudes are also affected by management. Does the quality manager have to use a correspondence store detective's disguise to gain access to the front office? Do quality control personnel get as many promotions as residents and alumni of the production department? Do performance evaluations give as much emphasis to quality as they do to production?

Inspection procedures are seldom controlled by the inspector. They are controlled by the quality department.

Quality characteristics relate to the product. There are, however, many ways to inspect a single item. Are inspectors encouraged to help select the tools and techniques that will be used?

Both production operators and quality control inspectors are ingenious at deviating from procedures when they have a "better way" of doing things. Those participating in the planning, however, are less likely to impose their wills on the system after the final decision is made.

DRAGON HEAD 3

Reliable, High-Quality Products Are Uneconomical

Try telling that to the Rolls Royce Company. If Rolls Royce quit producing a quality product, the company would go out of business. Rolls owners buy the car because of its quality. To them, price is secondary.

Makers of less expensive cars, however, still have tough quality requirements. Low price doesn't give them the license to be shoddy. Whether the product is toothpicks or space ships, it must perform in the manner intended, in the environment to which it will be exposed, for the time required.

DRAGON HEAD 4

Quality Is a Departmental Function

This head of the fearful dragon was conceived by the assumption that quality departments create quality like machines produce parts or magicians create illusions. Don't believe it.

The quality department's function is to help other departments coordinate their efforts to produce a quality product. Of course, the product must meet the standards, policies, and objectives set by top management.

The quality department's expertise lies in developing quality specifications, evaluating quality characteristics, measuring variations, and coordinating corrective action. A quality department's findings can be helpful if top management uses them. Otherwise, they have no more value than last week's cup of coffee, dehydrated by cigarette butts.

DRAGON HEAD 5

Quality Campaigns Ensure Quality

Using signs that urge workers to make "better quality" products can be effective if management doesn't stop there. These actions, however, are detrimental to morale and quality when they are not followed up with management action.

Is management doing all it can to give workers the necessary tools? Does management reward exceptional quality as much as it rewards outstanding production?

Workers quit listening to slogans the moment they differ from management's actions. This is particularly true in the field of quality. Workers do not strive for quality when volume of production is the only thing that counts toward their next promotion.

DRAGON HEAD 6

Defects Are Due to Lazy Workers

More defects are due to the attitude of managers than to the laziness of workers. Motivation is influenced by many things, and most of them are controlled by management.

How do workers think they are being treated? If people believe they are treated fairly, they normally try to do good work.

Does management skimp on training, raw materials, machinery, lighting, heating, cooling, and tender loving care? If so, workers assume that people in the trenches don't count, and neither does quality.

DRAGON HEAD 7

Good Design and Production Ensure Good Products

There is no question about the importance of good design and good production. Quality, however, doesn't end there. A product must be good enough to sell for a profit (marketing). The raw materials must be usable (purchasing). The product must be delivered in good condition (packaging and shipping). Without this support, the world's best design and manufacturing departments won't keep a company in business.

All departments in the company must be coordinated to produce a reliable, high-quality product—profitably. Only top management is in a position to direct and oversee the necessary coordination.

TIME OUT FOR HISTORY

Prior to World War II, Japan had a reputation for poor quality products. International competition required that they reverse this image, so they decided to slay the seven-headed dragon that spawns defects.

In order to improve their quality, the Japanese sought help from J. M. Juran, William Edwards Deming, and other quality control authorities from the United States. Our experts visited Japan and lectured to their business and industrial leaders. Our authorities spoke. Their leaders listened.

Since then, these same experts have lectured widely in the United States. They speak like ministers preaching to the choir. Those who really need to hear the message are seldom in the pews.

Now Japan has a better reputation for quality than the land of its tutors.

WHAT JAPAN DID RIGHT

The Japanese did two things that are giving them an edge.

First, the people at the very top of their business and industry listened to the U. S. experts. They recognized their responsibility for controlling quality within their organizations. They realized that control of quality starts at the very top and goes as far as top management insists, but no further. As a consequence, they ensured that their departments coordinated quality activities with each other. They also insisted that the agreements coming from these coordination sessions were implemented.

Second, the Japanese realized that specialized tools, techniques, devices, and procedures were needed to control quality. They mastered these tools, including scientific sampling, quality control charts, distribution curves, and other statistical methods. Now they understand how to evaluate quality and troubleshoot quality problems.

WHEN WILL THE UNITED STATES LEARN?

Many training courses in the broad field of quality control are available throughout the United States. These are generally attended by supervisors, engineers, and inspectors interested in professional growth. They are rarely attended by top management.

If the United States is to regain its position of world leadership in quality, top management must get involved. It must learn the language of quality. It can no longer delegate the entire quality function to subordinates.

Can executives manage a function without understanding its language? Isn't that expecting too much? It's like telling Aristotle to program an IBM computer three hours after bringing him into the twentieth century.

The ability of the United States to produce complex and sophisticated products is still outstanding, but this is not enough. The products must be safe, reliable, and strong enough to withstand abuse without hurting anyone. Customers must also feel they are getting their money's worth.

This book discusses the organizational principles and administrative procedures top management should follow to:

- Understand the language of quality control
- Recognize the tools of quality professionals
- Maximize product quality at minimum cost

Without this information, top management is recreating the Tower of Babel and wondering why nothing is ever built right.

2

Myths and Realities

Business has as many myths as an Athenian temple, and quality control is its high priest. Typical myths include the following.

MYTH 1

Quality Control Is a Necessary Evil

Reality: Quality control is necessary but it is not an evil, unless subjected to incompetent or disinterested management.

Many years ago, a midwestern aerospace company picked a production manager to be their director of quality assurance. He believed that quality control was an unnecessary evil. It took years to repair the damage he did during his short time in quality.

Prior to World War II, there were no quality departments as we know them today. They evolved rapidly during the war, as people began to realize that defective shells, tanks, and airplanes were costing lives.

As quality gained recognition, methods for evaluating and controlling this discipline were developed. These statistical methods helped the war effort so much that many of them were classified as confidential or secret.

After the war, many companies retained their quality departments; others formed new ones. Unfortunately, few of the newcomers understood the basic concepts. As a consequence, many of the early quality control departments added overhead without contributing to company profits. Some of these departments provided nothing more than "window dressing," giving quality control a bad name.

Quality departments with professional staffs and top management support, however, were able to reduce scrap, rework, customer complaints, and liability suits. They saved much more than they cost.

MYTH 2

Unanticipated Defects Are Inevitable

Reality: Most defects can be anticipated if you look for potential hazards.

How?

Don't assume customers will use your product properly. In real life, people pound with wrenches, pliers, knives, and shoe heels if a hammer isn't handy. They stand on chairs, barrels, boxes, and card tables if a ladder isn't close. The average workplace can turn into a Rube Goldberg nightmare if improvision isn't kept under control.

Product abuse is so widespread that manufacturers who don't anticipate it have unhappy customers. Twenty years ago, bridge table manufacturers wouldn't think of designing their product to support a 280-pound football player and his wife. Now, however, they must consider the possibility of having bulky customers use their product for reaching a light bulb. Think of the consequences, should the table collapse!

Never underestimate your customer's ability to think up cruel and unusual punishment for your product.

MYTH 3

The Law of Diminishing Returns Limits Quality

Reality: The law of diminishing returns seldom goes into effect until a reasonable effort has been made.

The law of diminishing returns states that the more time and money you put into an effort, the less the incremental gains. It is true that the last infinitesimal improvement towards perfection will cost an enormous amount. The problem, however, is that we usually quit too soon. We can come much closer to perfection, without excessive expense, if we plan properly.

The solution seldom lies in larger quality staffs or more inspection. It is usually found through better planning and improved controls.

MYTH 4

100 Percent Inspection Insures Quality

Reality: In many instances, 100 percent inspection is impossible.

As an example, you could never run ballistics tests on all of your solid propellant missiles. Once the test is run, there's no missile to sell. Tests such as these are destructive.

The solution is to inspect enough products to determine whether the manufacturing process is under control. If it is, then the quality of the tested units should approximate the quality of those that were not tested. If the inspected pieces meet the standard, the others will probably meet the standard. Probably is the key word, and statistics tell you what the probability is.

If the product can cause serious injury and the tests are not destructive, 100 percent inspection is probably wise. Such checking, however, must be carefully planned.

As an example, you want to be sure that the brakes on your new car have been tested before you take it for a drive. No one wants to detect defective brakes at a stop sign on a busy street with two police cars and an ambulance approaching the intersection.

Each product requirement is different. They range from defining an extremely hazardous flaw to describing a cosmetic requirement designed to promote sales. Where hazards are involved, every reasonable step should be taken to protect the customer.

Inspection of noncritical characteristics is usually controlled by economics rather than liability considerations. With noncritical characteristics, statistical methods and sound sampling procedures can help you improve profits by reducing scrap, rework, and customer complaints.

MYTH 5

Accepting Less Than Perfection Causes Mediocrity

Reality: If you make enough of anything some will be bad.

The important question is, "What are the consequences of a bad one?" If the quality characteristic is minor and few people care, the expense of reducing its frequency from one percent to .01 percent is seldom justified. There's no point in paying for a level of quality that customers don't want.

If a quality characteristic is important to the customer, do your best to ensure there are no defects, even if 100 percent inspection is required.

MYTH 6

Scientific Sampling Controls Quality

Reality: Scientific sampling helps determine whether material conforms to the quality requirements; it does not control quality.

When used, scientific sampling procedures must be followed rigorously. If necessary conditions are overlooked, "scientific sampling" becomes very unscientific.

Sampling does not guarantee the quality of your product. As production quality falls, the probability of it being accepted by a sampling plan also drops—but never reaches zero.

Attempting to control quality by catching "bad" lots with a sampling plan is expensive. In addition, the worse the production quality, the greater the probability that defective material will slip through the sampling plan and get shipped to the customer. This applies even when you sample and resample everything.

MYTH 7

Conformance to Design Ensures Quality

Reality: Conformance to design means that quality of the product will match quality of the design.

If the design does not anticipate all conditions under which the item will be used, consumers will probably pulverize it—and then blame you. Design engineers need to conduct adequate tests to ensure the product will perform as required, even though customer expectations may differ from those of the manufacturer.

Assume all of your customers have a Dennis the Menace in the house. Then, you should have happier customers and fewer liability suits.

MYTH 8

The More You Inspect the Better

Reality: This is not true unless you take the samples properly, inspect the right thing, in the right way, with the right equipment, using the right procedures. In addition, you must pick the right time, the right place, and interpret the results in the right way.

As an example, if you want to determine the average annual temperature at a specific location, don't take all the readings at noon. You could collect readings for a thousand years without coming up with the right average. It would always be too high, since noon temperatures are close to the highs of the day. This example may seem oversimplified but similar errors have happened in industry.

Diminishing returns must also be considered. If you have inspected 5 items, looking at 5 more improves your evaluation. If you have checked 50, looking at 5 more may improve your evaluation, but not as much.

MYTH 9

Large Quality Staffs Ensure Good Quality

Reality: Large quality staffs may become the problem rather than the solution.

One West Coast aerospace company studied factors influencing defect rates. The variable having the most effect was the work load of their quality control inspectors. The more time the inspectors had to look for defects the more they found. Many inspectors felt they had to justify their existence. They looked for unimportant, cosmetic problems when they had free time.

Assuming that all quality problems can be solved by enlarging your quality staff is like trying to inflate a flat tire by putting more gasoline in your tank.

When quality problems arise, your first question should be, "Why?" Once this question is answered, you can decide what to do about it. Your staff may need additional training. Your quality department may be poorly organized. Your tooling and equipment may be inadequate.

Some managers feel their quality departments are totally responsible for product quality, but their quality staff doesn't manufacture a single item.

Blaming your quality department for poor quality is like blaming your secretary because you forgot your spouse's birthday.

MYTH 10

Good Public Relations Ensure a Good Public Image

Reality: Good public relations will help your image if you have a good product. If your quality falls, so will your public image.

Many years ago, a large brewery cut production costs at the expense of quality. They were sure the public didn't know the difference. They thought they could do more for sales and profits by increasing their advertising budget. That company is no longer an independent brewery.

One of the most common slogans in advertising is "Quality products since _____ ." The blank is usually filled in with the year the company was founded. The farther back in time, the more meaning there is in the slogan. Any company that survives for more than 50 years must have good quality control, even though they may call it by a different name.

A good public relations program attracts customers. If performance doesn't match promise, however, few customers come back a second time. Repeat sales require good quality at competitive prices.

MYTH 11

It's Cheaper to Let Customers Find Your Defects

Reality: When defects reach the customer, repeat business suffers. Most companies that depend on customers to find their defects go out of business.

MYTH 12

Quality-Minded Workers Always Make Good Products

Reality: Quality-minded workers will produce the quality inherent in:

- The engineering design
- The raw materials
- The capability of the production equipment
- The working conditions
- The worker's training and instruction

Quality-minded workers follow instructions and adhere to training guidelines. There are, however, so many variables they can't control that having quality-minded workers, by itself, won't ensure a quality product.

MYTH 13

Top Management Doesn't Have to Understand Quality Control

Reality: Without the understanding of top management, quality control becomes little more than a slogan.

Many executives avoid the quality function, giving their subordinates full responsibility for product quality, but without commensurate authority. Rather than trying to understand the subject, they pretend it doesn't exist.

An executive's exposure to the technical aspects of quality control is similar to a mouse's exposure to a hungry python. If they fail to learn the strengths, weaknesses, and language of the enemy, they become victims of their own ignorance. In time, however, quality control can become the executive's friend.

In Japan, it has been proven that the technical aspects of quality control can be mastered by management. Japanese executives have learned the language. They recognize the strengths and work on the weaknesses of their quality control departments.

Are we interested in doing as well?

3

Keys to Managing Quality

Five keys to managing product quality are:

1. Knowledge of customer requirements
2. Anticipation of defects
3. Detection of defects
4. Coordination
5. Effective problem solving

KNOWLEDGE OF CUSTOMER REQUIREMENTS

Until the industrial revolution, artisans were responsible for their own quality. If they sold junk to important people, the retribution was swift and sure. It didn't take artisans long to anticipate customer requirements when the price of failure was a tour of the wealthy buyer's dungeon. Judgments concerning quality were subjective, however, forcing workers to make conservative decisions.

Customer requirements became more objective once people started asking for items that were exactly alike. This gave them a standard to work with. It was necessary, however, to determine which characteristics were important. As an example, one nobleman might want a shoe just like the king's. If his foot was twice as big, however, "just like" could not include size.

Evaluation of customer requirements became more objective when Eli Whitney implemented the concept of mass production and interchangeable parts. Whitney had each worker produce a single component to prescribed specifications. This procedure improved efficiency. It also proved that quality could be measured and parts could be interchanged.

Eli Whitney's rifles were assembled from parts chosen at random. This advancement proved handy in battle where good parts from two nonfunctional rifles could be used to make a good one—assuming the rifles had different defects. As a consequence, a new customer requirement was born: Rifle parts had to be interchangeable.

Mass production eliminated the inefficient process of making parts, trying them out, and making adjustments until they fit. Mass production also separated the worker from the customer. It became more difficult to tell if customer requirements were met.

Today, customers require quality and reliability commensurate with the price they pay. When they buy a Lincoln, they expect Lincoln quality and reliability. They don't expect squeaks and they don't expect to rebuild the engine after 50,000 miles. When they pay for a Ford, they expect Ford quality and reliability. They will put up with a few squeaks, but they don't expect the car to disintegrate the day its warranty expires.

Lincoln parts are usually more precise and more expensive. Ford parts may have looser tolerances but even those requirements must be met. The alternative is excessive scrap, rework, and customer complaints. An astute tradeoff between tight specifications and reasonable price often determines whether customers will buy the product.

ANTICIPATION OF DEFECTS

Anticipation of defects is enhanced by an understanding of variation. No two Jeeps, Chryslers, Continentals, or Rolls Royces are exactly the same. Variation is a fact of life—and the cause of most defects.

Early in the movement towards mass production, people found that some "interchangeable" parts fit but others required rework. It soon became evident that they needed to limit the variation in components that were supposed to be interchangeable. As a consequence, specification tolerances were developed.

Most manufactured products have upper and lower specification limits. As an example, specification limits for the length of a large ballistic missile chamber might be 154.800 inches on the low side and 155.200 inches on the high side. Chambers falling within these limits would have acceptable lengths. The rest would be defective. They could be reworked to meet the specifications, discarded, or accepted for further processing.

Astute managers review their specifications whenever a significant number of defects are accepted for further processing. If the "defective" parts have neither hurt production nor caused customer complaints, these managers change the specification.

What is the nature of variation?

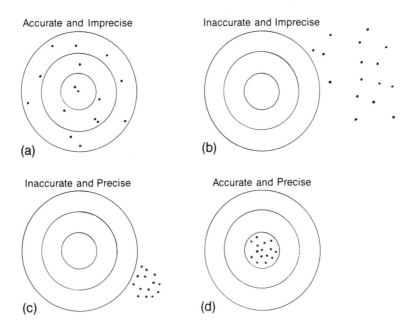

Accurate and Imprecise   Inaccurate and Imprecise

(a)   (b)

Inaccurate and Precise   Accurate and Precise

(c)   (d)

Figure 3.1 The nature of variation.

Figure 3.1 explains two types of variation. Suppose you were on the rifle range. If your target looked like Figure 3.1a, your shooting would be accurate but imprecise. You were accurate because your "average" shot was in the bull's-eye. You were imprecise because all but two of your shots missed the bull's-eye. In other words, you had too much scatter.

If your target looked like Figure 3.1b, your shooting would be inaccurate and imprecise. Your shots had excessive scatter, and all of them were to the right of the bull's-eye.

If your target looked like Figure 3.1c, your shooting would be inaccurate but precise. Your shots had little scatter, but they all missed the bull's-eye.

If your target looked like Figure 3.1d, your shooting would be accurate and precise. All of your shots were within the bull's-eye.

In industry, variation is similar. If your process is accurate but imprecise, you need to review your process capability and your specifications. If your specifications can't be relaxed without causing production problems or customer complaints, you need to improve your process capability. This could involve new equipment, different raw materials, or improved procedures.

If your process is inaccurate but precise, you probably need to change one or more process variables such as settings, speeds, temperatures, or pressures.

If your process is inaccurate and imprecise, you need to react to both types of problems.

Precision is a matter of scatter. Accuracy is a matter of how close the central point of the scatter is to the bull's-eye. You can have either, neither, or both.

When making general purpose ash trays for skid row bistros, you have lots of leeway with the precision and accuracy of your measurements. When shooting the cigarette from the mouth of an attractive model, you have very little leeway.

It is important to know how tight specifications have to be. A West Coast aerospace engineer once needed a container, the size of a small ashtray, for curing polymer. He explained the purpose to the expediter but failed to explain that he wanted the dimensions accurate to a quarter of an inch. Three weeks later, the engineer received a highly polished, precise, and accurate tray that met the nominal dimensions on his sketch within .001 inch. Needless to say, the tray's cost exceeded the engineer's estimate.

DETECTION OF DEFECTS

During the early stages of the industrial revolution, measuring instruments were crude and disputes over measurements were common. As a consequence, it was difficult to separate defectives from acceptable parts.

To alleviate this problem, a Swedish engineer named Johansson developed a set of "perfect" standards for length. These standards, which helped bring order out of chaos, are now called Jo blocks.

Jo blocks, and other measurement standards, go a long way toward helping detect defects. Even these standards, however, are not perfect. Every measurement has some error, although its magnitude may be less than a millionth of an inch. There are no perfect measurement standards, no perfect production tools, and no perfect measuring devices. Close, maybe, but not perfect. If two items were perfectly identical, you couldn't prove it because measuring devices are not perfect.

In order to detect defects with confidence, you need accurate and precise measuring equipment. Without it you will reject too many parts that are good and accept too many parts that are bad.

COORDINATION

Coordination is one of the most neglected keys to effective quality control. Most people assume that quality will be fine as long as each department does its job. Without coordination, however, a problem-solving exercise resembles a Keystone Cop comedy, with everyone rushing off in different directions.

A common alternative to coordination is giving autonomy to each department in the company.

Is this wise?

- Should the marketing department determine what the company will sell without input from those who make the product?
- Should the purchasing department select vendors without consulting those who test incoming materials?

- Should the engineering department design the product without input from sales or marketing?

Coordination of the quality assurance effort requires a knowledge of process monitoring, defect identification, statistical trouble-shooting, and the evaluation of customer feedback. Although the head of quality assurance normally understands these functions, knowledge is not enough. The coordinator must also have the status necessary for getting the job done. In the army, generals seldom take direction from captains. In industry, vice presidents seldom take direction from department heads.

Who should be the coordinator?

If the head of quality assurance is a vice president and has the skills previously mentioned, that individual is a natural choice. If the head of quality assurance is a recent castoff from production or some other division, choose someone else.

The chief coordinator needs clout and status equal to the heads of production, marketing, and engineering, plus a knowledge of the statistical tools to be used. If such a person can't be found, train or hire someone. If this does not sound politically or economically reasonable, admit you are not interested in quality and concentrate on price. Good marketing divisions can get away with selling junk—if it is cheap enough.

EFFECTIVE PROBLEM SOLVING

Before the industrial revolution, master craftsmen and their apprentices converted raw materials into finished products. Very little was done without involving the master craftsmen who were responsible for quality.

Today we have mass production. Thousands of people can be working in a single plant where supervisors and engineers are unaware of production problems for hours or even days. Managers may not hear of these problems for weeks.

The operators are now the "master craftsmen." No one else really knows the details of what's going on in the shop.

In order to capitalize on the operator's closeness to the work, many companies are training shop personnel in statistical techniques. This training helps them participate in the correction of quality problems.

In Japan, the training goes one step further. The Japanese even teach executives to use statistical tools. This training helps upper-level managers to participate in the correction of quality problems. When they can contribute, they don't get in the way as much.

Steps used in effective problem solving include:

- Gathering data
- Tabulating
- Summarizing
- Analyzing
- Interpreting

The recording and gathering of data is normally performed by operators. If they know how the data will be used, they will be more conscientious about recording and gathering it properly.

During World War II, weather observers in the Fourth Weather Squadron frequently showed zeros as the last digit of their pressure readings. The second most common number was five. Theoretically, 10 percent of the end digits should be zero, 10 percent should be 1, 10 percent should be 2, and so forth. Once this discrepancy was pointed out by their headquarters, the observers took more care in reading and recording the last digits. As long as no one complained, the observers assumed their readings were acceptable.

Operators must know what is needed, how it will be used, and how much accuracy is required. This knowledge should be provided before they unintentionally give you misleading information.

Tabulation is a statistical technique that makes large quantities of data readable. This task is usually assigned to the quality control department and requires some exposure to statistics.

Summarization involves boiling down the tabulated data so that it can be used easily. In most cases, summarization is also handled by quality control personnel with some exposure to statistics.

Analyzing data is the most complex part of statistical trouble-shooting and requires additional training. The detail work is usually performed under the direction of statisticians. The current trend, however, is to involve plant operators. Once they are trained in the analysis of data, they can see the importance of numbers they generate.

When managers are given training in the analysis of data, they can see the source of analytical errors and know what to look for when results are presented to them. They are also better prepared to communicate with their statisticians and engineers.

Statisticians are people, despite rumors to the contrary. When statisticians are not given direction in terms they understand, they work on projects they consider important. When given direction in terms they understand, they work on projects their managers consider important.

Interpreting statistical analyses requires an understanding of the purposes, advantages, limitations, and potential benefits of statistical studies. With a good background in these areas, managers can ask the right questions when statistical data are presented. Without this background, many of them bury their heads in the scrap pile and reject all analyses. Some are more comfortable hiding from new ideas than going through the trauma of showing their ignorance to subordinates. They are like the few accountants who refuse to computerize accounting systems because they don't understand computer programming.

Tools, used in statistical trouble-shooting, include:

- Basic statistics
- Process capability studies
- Tests for comparing processes and products
- Frequency distribution charts
- Control charts
- Precontrol
- Regression analyses
- Designed experiments

Although it is seldom necessary for operators or managers to perform statistical tests, they should be aware of their advantages, disadvantages, and limitations. These tests will be covered in subsequent chapters.

4

Key 1: Knowledge of Customer Requirements

Who are your customers? Men? Women? Senior citizens? Teenagers? Southerners? Easterners? The underprivileged? The affluent? Each group has different expectations. As an example, the underprivileged usually stress price more than the affluent. Senior citizens usually stress safety more than teenagers.

Marketing departments normally conduct surveys before committing their companies to a new product. Based on these surveys, the marketing staff makes assumptions about who will buy the product.

Additional studies are conducted to see if similar products are being sold by other companies. If competition is already entrenched, marketing develops plans to ensure that their product will attract customers away from the competition. The ideal marketing survey also determines how long customers expect the product to last and what defects will be tolerated.

One California home builder didn't bother with marketing surveys. He knew what kind of houses the area needed and he built them. His houses were solid and functional but he had few models and out-of-vogue designs. Within three years, he was out of business. He was too stubborn to admit that buyers might not want what he "knew" they needed.

The owners of a jewelry firm had a similar problem but reacted differently. The jewelers thought their customers would object to minor blemishes in one of their highly engraved products. They removed the visual defects by buffing them out. As an afterthought, they submitted buffed and unbuffed samples to a panel of potential customers. The panel voted for those without buffing.

The customers didn't care about the minor visual defects the buffering removed; they were more interested in the sharpness of the engraving, which buffing destroyed. Thanks to the test, the jewelers were able to reduce production costs by eliminating the buffing. The customers also benefited by getting the sharp, clear engraving they wanted.

One thing customers normally want is reliability. In a technical sense, reliability is the likelihood that the product will perform in the manner intended, in the environment specified, and for the time required. To most people, this is the same as quality. Customers don't want a product to disintegrate the day after the warranty expires; they want it to last forever.

Customers object to being inconvenienced by someone else's sloppy work. Getting the product fixed is one inconvenience that bothers everyone, even if they don't pay for the repairs. With inexpensive items, customers often ignore the problem or throw away the item. Then they hold a grudge against the company and quit buying its products. In most instances, companies never find out why customers defect.

In the early days of quality control, attention was focused on quality characteristics which could be evaluated during production or at the end of the manufacturing process. Performance testing, if there was any, was "quick and dirty." The tests were run at plant conditions for a short time. Some companies tested to determine the probable life of the product, but few considered the environmental conditions under which the product would be used. As a result, the companies received a rash of complaints they couldn't understand.

With the advent of the space age, product life and environmental testing became important. This trend toward reliability soon spilled over into the consumer market. Now customers don't consider a product reliable unless it is usable when it goes out of style. They'd rather not have so many problems that they get to know repair shop supervisors on a first-name basis.

For years, reliability evaluations were subjective. Since then, reliability engineers have added the concept of probability.

What is probability? Probability deals with the likelihood of something happening; it is the concept that enables Las Vegas casinos to make money. Many advances in the theory of probability came from gamblers who wanted to improve their winnings. As an example, blackjack experts found that knowing the probability of getting a face card gave them an advantage over neophytes who relied on nothing but hunches.

You might want to know, "What is the probability that you will make money in Las Vegas?" You can be assured that it is less than 50 percent. If it were higher, the house would go broke.

The probability that you will get heads if you flip a coin is 50 percent. Heads should come up half the time unless the coin's weight is unevenly distributed. When you gamble with dice, the probability of getting a six with a single dice is one out of six unless the dice are loaded.

If you own a TV set, the probability of it needing repairs is more complex.

Probability of failure depends on:

- Complexity of the product
- Nature of its components
- Environmental conditions
- Stress applied

Complex products, like TV sets, are more likely to fail than simple products, like hammers. Suppose each critical component in a TV set has a probability of failure during warranty of only .001 or one in a thousand. This sounds like good protection against customer complaints. However, if there are 50 similar parts and the failure of any one could cause the TV to stop working, the probability of failure increases to 5 percent or 1 chance in 20. A company that could survive one customer return in 1000 sales might not be able to survive 1 in 20.

In addition to the complexity of the product, there may be considerable variation in the nature of the components. Metal components, as an example, are much less likely to fail than their average plastic counterparts. Many plastics are brittle, weak, and easily scratched.

Environmental conditions are often overlooked, even today. Ask the design engineers: "What will happen to the product at high or low humidity, extreme temperatures, or unusual atmospheric pressures? What will happen if the product gets exposed to moisture, salt air, sunlight, or electrical fields?" These hazards can occur alone or in combination with one another.

Many of the first home computers broke down in less than a year because of corroded contacts. Users spent many hours carrying their computers to repair shops until they learned how to correct the problem by cleaning the mating surfaces. Eventually the manufacturers got the message and made contacts that resisted corrosion.

Manufacturers have also found that they must design their products to withstand unusual stresses. As an example, automobile brakes might work fine in normal use. If used continuously, some brakes overheat and turn spongy. Even though inexperienced drivers may use their brakes instead of compression when coming down a long grade, they still expect good brakes when they get to the bottom. If the brakes can't stand this kind of stress, the company may have a few dead customers.

Brake designers should anticipate the stresses resulting from inexperienced drivers riding their brakes as they come down from the mountains. The world is full of greenhorn drivers and steep mountain roads.

When designing and testing a product, it is necessary to consider what happens when your customer subjects it to unusual stress. This includes using it frequently at short intervals or continuously over long intervals.

As an example, it causes poor customer relations when you design home computers to operate for fewer than 2 hours per day. With most users, 2 hours per day is adequate. Some, however, use their computers 12 hours per day, 300 days per year. Customers who give computers this stressful treatment are the ones most likely to complain if the computer doesn't stand up.

Fifty years ago, the legal precedent was, "Let the buyer beware." Now the precendent is, "Let the designer prepare."

HOW TO IMPROVE PRODUCT RELIABILITY

You can design reliability into a product by developing systems that are either life-safe or fail-safe.

Life-safe systems reduce the probability of failure by including large safety margins in the design. As an example, if a part is expected to carry a maximum load of 100 pounds before failure, you might design it to carry a load of 120 pounds. If the part lasted 1000 days when designed to handle the maximum expected load, it might last 10,000 days if designed to handle the extra 20 pounds.

One manufacturer of heavy equipment for construction and road work had a reputation for overdesigning his vehicles; they seldom failed. This reputation was maintained until he sold out. Although he has not been in business for over a decade, many of his vehicles are still in service despite the severe stresses involved in construction work.

The concept of fail-safe goes back to the chemical and petroleum industries prior to 1950. These industries used control valves, which regulate the flow of gases and liquids. Many of the valves are opened and closed by instrument air that comes from a compressor. Designers and startup engineers are concerned about what will happen if the air compressor shuts down, stopping the flow of air to the control valves.

The question they faced is, "What is the safe position of the valve?" If the safe position is closed, then they design their control valves to close when the instrument air is off. If the safe position is open, they design the valves to open when the instrument air is lost.

Control valves regulating the flow of gas to a steam boiler are good examples. In case of emergency, the designers want the gas to be turned off. If the flame of the boiler goes out and the instrument air also goes off during an emergency, resumption of gas flow into the hot boiler might start an explosion.

The absence of gas is the safe situation, so the designers want the valves to be closed when the instrument air fails. They also want assurance that the gas flow doesn't resume when the instrument-air compressor starts up again.

A fail-safe condition can also be achieved through redundancy; back-up devices can be designed to take over if the main system fails. As an example, hydraulic brakes on an automobile may have a cable linkage that is slack until the hydraulic system fails. Then the backup cable system can take over and stop the car.

Either life-safe or fail-safe systems may increase costs. These must be balanced against the consequences of failure. Where failure can result in loss of human life and occasional failures are anticipated, the cost of prevention is normally warranted.

Designing reliability into a product is only the first step. Tests are also required to ensure that reliability is actually there.

A tea company once formulated and tested a new product line during the fall and winter. All went well until the following summer when they received a rash of customer complaints: The tea didn't taste right. An investigation showed that the change in taste came from variability in the weight of the tea packages. Too much tea resulted in the flavor being too strong. Too little tea resulted in it being too weak.

Variability in tea weights was traced to the tea balling up, or forming balls of tea, during processing. Balling of the tea was traced to the summer's high humidity. The company had to reformulate the tea. Now they include humidity tests as part of their qualification process.

Field tests are also becoming popular. Pilot programs are used where the product is test marketed in a limited area to see how it will sell. Test marketing also helps identify major flaws before the product is distributed nationwide.

Many new beers, for example, are tried out in select cities before they are given national distribution. The breweries want to be sure the product will sell before they commit themselves to creating a huge inventory. Customers don't always buy what they say they want when filling out market surveys.

Aging is another critical problem. Food and chemical products spend time in stores before they are sold. Most of them also spend time on the customer's shelves before they are used. If a product is expected to be good for several years and goes bad in less than one, the manufacturer has a problem. It is best to know about it before the product finds its way to a million homes across the country. Product recalls are expensive.

Accelerated aging tests involve exposing the product to temperatures and humidities more severe than those it should see in the field. In fewer than six months, a well-designed test can simulate two years of storage at normal conditions.

Don't test the loyalty of your customers. Do preproduction testing instead.

5

What Bugs Customers?

Defects in consumer products are as diverse and prolific as snowflakes; no two are exactly alike and they just keep coming. Typical, real life examples include:

1. The fan on a 1200-watt, electric, forced-air heater fell off the shaft after one year. The fan was held in place by a nut-and-lock washer that vibrated loose. The designers could have keyed the fan to the shaft and held it in place with a snap lock, but they didn't.
2. The plastic on an automobile steering wheel shrank, developing four 360° cracks. A steel core saved the day. The wheel, however, looked like it came from the junk yard after standing in the rain, snow, heat, and cold for the last 20 years. Aging tests, under adverse weather conditions, could have detected this design weakness before the car went into production. Then, the problem could have been corrected by changing the plastic to a material that didn't shrink or crack when exposed to harsh weather.
3. Three tubes of dental cream split soon after purchase. The tubes had severe longitudinal scratches that caused them to fail after a few squeezes. As a consequence, that manufacturer lost at least one customer and probably more. People love to talk about product failures as much as they enjoy gossiping about

the next-door neighbors. With good quality control procedures the company could have corrected the problem and kept its name from being emotionally discussed at many bridge and sewing circles.

4. A well-known hand-held calculator malfunctioned almost immediately after purchase. It had faulty switches under the data entry keys. A new calculator, sent under the quarantee, had the same defect. The replacement went into the trash can along with the manufacturer's reputation.

5. An aluminum piston in a new automobile broke up after 20,000 miles of travel. Metallurgical examination showed it had a casting or forming crack which must have occurred during manufacturing. If the crack had been detected by in-process inspection the company's reputation would have had at least one less blemish.

6. An electronic watch was built into an expensive ballpoint pen. Within a week of purchase, the viewing window fell out. A second watch-and-pen combination had the same defect. The window could have been designed with a physical restraint to prevent it from coming loose, but it wasn't.

7. A cassette adapter for an automobile tape deck malfunctioned after a year. One of the wires had broken due to normal flexing. Accelerated aging tests, prior to production, could have identified this deficiency. Then a more suitable wire could have been incorporated into the design.

8. An electric clock that projected time of day on the ceiling became excessively noisy after one year. The mechanism was sealed and could not be reached without forced entry. Problems with the clock could have been detected by preproduction aging tests. Then the design could have been changed to provide better lubrication for the noisy parts.

9. An expensive calculator stopped working while under warranty. It was sent back to the factory where they took four months to replace it. The poor response time was as irritating as the product failure. When production problems occur, good customer service helps minimize the impact of occasional lemons that slip through the system.

10. One of the reception bands of a multiband radio went dead two weeks after purchase. It was under warranty and was replaced, but the second radio had poor sensitivity. Customer service did a poor job ensuring that the replacement unit was better than the one that came back. Faulty replacement units make customers wonder whether companies can do anything right.

11. The recording button on a tape recorder slipped out of position just after the warranty expired. It took two months to get it repaired and the repairs cost about half the purchase price. A year and a half later the problem recurred. This time, the recorder was junked and replaced with a different brand. Aging tests, prior to production, could have identified the problem and enabled the manufacturer to improve the design.

12. The rear window of a new car was trimmed with a plastic strip, the ends of which were butted and covered by a short metal sleeve. In a few months, the plastic had shrunk until the metal sleeve fell off, leaving an obvious gap. Aging tests, prior to production, could have identified the problem. Then the design staff could have corrected it.

13. A new car contained an electronic tachometer which went bad while under warranty. Two months later, the replacement failed. By then, the warranty had expired; a second replacement would cost about $100. Rather than get the tachometer fixed again, the owner used it as a conversation piece. Whenever he had guests, he gave them a long explanation of how the automobile company didn't know how to provide tachometers that work. Even purchased parts should be given aging tests prior to production.

14. A tape deck stopped producing sound after three years—a wire from the sensing head had broken due to flexing. Inadequate preproduction testing was apparently at fault.

All of these defects were identified by one person. Other people probably have similar lists.

Why were these specific defects remembered? Customers seldom forget problems that:

- Cause an inconvenience
- Become repetitive
- Reflect disinterest on the part of the company

Customers seldom get compensation for the time, effort, postage, gasoline, shoe leather, and heartburn medicine consumed in their quest for replacements, even when covered by a warranty. No one wants to spend time and money to correct another's mistakes. Most of us also object to doing without the defective item while it is being replaced. This shortcoming is less annoying when a loaner is furnished but loaners are expensive; few companies provide them any more.

Repetitive problems are even more frustrating. Companies that don't ensure their replacements are functional invite competition into their territory. Very few customers will put up with the inconvenience of requesting a second replacement without writing the company off. They consider it a born loser, tripping over its shoelaces on the way to Chapter 11 bankruptcy.

A company shows it doesn't care when it replaces defective merchandise with something as bad or worse. It also shows disinterest when it fills its customer service organization with uncaring people. Don't staff this important part of the company with rejects who were dumped by other departments because they couldn't perform.

Failure to send replacement parts by return mail is like telling customers they don't count—even thought they're irreplaceable. When you run out of customers, you run out of income. Only expenses last forever.

6

The Ultimate in Displeased Customers

Mix an injured customer, an opportunistic lawyer, and a consumer-oriented jury and you can have an adverse liability settlement that could drive you out of business. The ultimate in displeased customers is the one who sues.

Some of our grandparents dreamed of their ships coming in. Others dreamed of inheriting riches from long-lost relatives. Now many dream of losing relatives in an accident and suing all possible contributors for millions. Unfortunately, the biggest slice of the pie will probably go to the lawyers.

In the old days, lawyers who took advantage of accidents were called ambulance chasers. Now they are called consumer advocates. They advertise on TV encouraging people to sue, even though the injured party may have been negligent. Consumer advocates don't chase ambulances any more. They advertise.

Consumer-oriented juries are awarding larger and larger settlements, even though the companies may not have been at fault. The popular notion today is: "If you want to get rich, sue."

Paul J. Grant, Esq., a member of the Michigan Bar Association, speaks of one such case in the September 1985 issue of **Quality** magazine. A multimillion-dollar verdict was awarded to a consumer where there was almost no technical evidence to support the existence of a defect. The jurors pitied the plaintiff and believed the company could afford the judgment.

34

Today we live by a new set of rules. Traditional defenses to liability suits are much less effective than they used to be. Typical defenses of the past included:

- Obvious hazards: "The customer should have been more careful."
- Warnings: "We warned the customer—that relieves us from responsibility."
- Nonhazardous products: "Our product couldn't hurt anyone."
- Customer negligence: "The customer misused our product."
- Compliance with sample plans: "We complied with Military Standard 105D."

OBVIOUS HAZARDS

Some products are obviously hazardous. Examples include knives, motorcycles, gasoline, chain saws, grinding wheels, and blasting fuses. If handled with reasonable care they do not present a problem. If the customer uses them carelessly, few juries will sustain a suit against the manufacturer.

Warnings and precautions, however, are still necessary. Anyone shipping explosives without labeling them properly is asking for trouble.

Other precautions are also helpful. One manufacturer, who has been making blasting fuses for over one hundred years, is sued every year. Injured miners claim the fuses burned too fast and they were unable to get away from the blast area.

The company always invites the plaintiffs to come into the plant and select their own sample of fuses. They are then asked to ignite and time them in court. No fast-burning fuse has ever been found and the company has never lost a case. The evidence indicates that the miners did not give themselves enough time to get away. The company protected itself by having an excellent design and by exercising tight controls over its manufacturing processes.

With hazardous products, such as blasting caps and fuses, companies have at least four ways of avoiding liability judgments:

1. They can design the product so it cannot malfunction.
2. They can test every unit produced. (This approach can't be accomplished if the test is destructive. Enough units must be tested, however, to establish confidence that production meets the design standard.)
3. They can record inspection and test results to provide evidence of high quality.
4. They can let plaintiffs select any item of production and test it in court.

The courts have made it clear that it is the manufacturer's duty to conduct tests to assure a safe product. If any hazardous defect is discovered, the company must react immediately to prevent recurrence, even if the problem was never seen before. A single hazardous defect must be a red alert for the company.

WARNING

During the youth of our grandparents, "let the buyer beware" was a common defense. Now the producer must beware. Warning messages, hidden in voluminous user instructions, may or may not help manufacturers defend themselves in court.

How many times have you read all of the instructions before starting a new lawn mower? The lawn could grow another inch before you finished reading.

In addition, how many lawn mowers are bought by people who can't read English?

Warnings that can't be read without a microscope will seldom protect the manufacturer from a liability suit. Warnings written in large print on the product itself are more likely to sway a jury. When the product is sold in a bilingual area such as New Mexico, the warnings should also be bilingual, especially when you can expect the product to be used by new immigrants or transient workers.

NONHAZARDOUS PRODUCTS

Limiting production to totally nonhazardous products may not protect you from liability suits. As an example: "How can anyone get hurt with an ashtray?"

That's what an automobile manufacturer thought. The company classified their dashboard ashtrays as nonsafety-related items. They were inspected by a sampling plan.

A passenger in the front seat of one of their cars was smoking and pulled out the ashtray. There was an accident and the passenger was thrown against the dashboard. Edges of the ashtray were ragged and sharp due to a defect in manufacturing; had the tray been inspected it would have been rejected. The passenger's leg hit the open tray, causing severe cuts and extensive damage, so the passenger sued the automobile company.

The injured passenger's attorney claimed that the ashtray was installed in the car because a sampling plan was used. The court was convinced that the company chose sampling for economic reasons. They also assumed that 100 percent inspection would have kept the defective ashtray out of the car. In the end the automobile company was held liable; the passenger won the case.

This example shows that there are no non-safey-related items. Virtually everything can become dangerous if it is defective enough. Manufacturers must try to anticipate unsafe conditions no matter what they make.

CONSUMER NEGLIGENCE

At one time, proof of customer negligence was a sure defense. Consumer-oriented courts and juries, however, have changed the rules. Million dollar awards have been made even though the court considered the customer to be 80 percent responsible for the injury.

The line between intended use and abuse can be very thin. Slightly overstressing a product may have little or no harmful effect. Some of this must be anticipated. On the other hand, using an ordinary drinking glass as a hammer is clearly unwarranted.

Technically, using an item in a novel way, not intended by the manufacturer, is misuse of the product. If such abuse can be anticipated, however, the manufacturer is obligated to warn the customers. This is especailly true when the risk of injury is involved.

Some abuse is inevitable. As an example, instructions with most electronic calculators warn the user against rough handling. A large variety of calculators, however, can be carried in a shirt pocket. When the user leans over, the contents of the pocket often fall out. If a calculator falls 4 feet, along with pencils and pocket secretaries, it will be jarred.

Calculator manufacturers, who don't design their equipment to withstand these common falls, may lose customers. Technically, the dropped calculators have been abused. Some calculators, however, will inevitably be dropped and manufacturers should design their product accordingly.

Many warranties have disclaimers that attempt to free the company from responsibility when there is abuse by the customer. Customers, however, tend to excuse themselves and blame the manufacturer for failures. This is accentuated when customers find competing products that withstand similar abuse. Companies that rely too much on disclaimers lose customers.

SAMPLE PLANS

Compliance with recognized sample plans like Military Standard 105D is another risky defense in court. The case of the defective ashtray is one example.

Part of the public's aversion to sample plans comes from the fact that many plans permit acceptance of a shipment even though one or more units in the sample is defective.

As an example, suppose you have a shipment of 1000 parts and are using an acceptable quality level (AQL) of 2.5 percent. Using Military Standard 105D you would normally base your acceptance or rejection on sample plan "J." This plan allows you to accept the shipment even though you find 5 defective parts in an 80-part sample.

This sample plan is more stringent than many used by industry. Yet, if you use this plan and receive a shipment that is 10 percent defective, you will accept the material 20 percent of the time.

PROTECTING YOUR COMPANY

Some alert companies use quality teams to reduce their vulnerability. These teams are formed before the first protopype of a new product is is built.

Departments that should be represented on the teams include design, quality, manufacturing, field services, marketing, purchasing, packaging, and shipping. Quality team assignments should include:

1. Conduct a design review to find flaws in the drawings and specifications.
2. Ask: What can go wrong? How can the customer misuse the product? What environments can the product be subjected to?
3. Attach a form to each drawing listing the two most likely failure points for each part or assembly.
4. Conduct failure analyses.
5. Determine which quality characteristics are critical and which have an impact on user safety.
6. Determine what happens if any characteristic is defective.
7. Analyze a fault tree that identifies undesired events. For each undesired event identify the conditions that might cause it. These conditions can be considered individually. They should also be considered in groups where several adverse conditions are present at the same time.
8. Review probable failure rates for components and assemblies.
9. Justify excluding parts and assemblies not on the list.
10. Make prototypes. Then arrange the parts into three piles: those that will not cause failure of the final product, those that might cause a failure, and those most likely to cause a failure.
11. Run tests on the prototypes from the last two categories. Include all possible modes of failure.
12. Take at least three prototypes and literally "beat them to death." This should include exposure to "cruel and unusual" environments and severe overstressing. Continue the tests until the prototypes fail; this helps identify weak points in the design before the product goes into production.
13. Determine what inspection and testing should be conducted on incoming materials.
14. Determine what records should be kept to protect the company in case of liability suits.

The cost of the above procedures will vary depending on the nature of the product and how thoroughly the procedures are carried out. The tests are expensive but the alternatives could be catastrophic.

As an example, consider the compact cars that frequently caught fire when rear-ended. The company recalled over a million of them for fuel tank modifications. The federal government declared that the fuel systems on these models were dangerously susceptible to gas leaks and explosions when struck from behind. Over 50 people died in the resulting fires.

The company never conceded that there was a safety problem. To dispel public concern, however, they agreed to recall the cars and add protection to the gas tanks. Had this been a part of the original design, many of the accident victims might be alive today.

Another company's steel-belted radial tire provides a second example. During 1978, the government stated that the tires had a "safety-related defect" which led to thousands of tire failures. The manufacturer claimed that users overinflated the tires causing the problem. The company, however, took the tire off the market; those in service were recalled. Who's responsible?

Automobiles have been rear-ending each other ever since the vehicle was invented. This is not an unpredictable event. Therefore, automobile manufacturers should design and build their cars to minimize the hazard of fire following a crash.

Tire life and performance has greatly improved over recent decades. Blow-outs and tread separations, however, have created problems ever since pneumatic tires hit the freeways. Blow-outs and tread separations will happen. Shouldn't tire manufacturers be expected to design and build their products to minimize the risk of these predictable accidents?

Having customers find your defects is bad enough. Expecting them to forfeit their lives in the process is going too far.

If hazardous failures can be predicted, designers should predict them and beef up their designs accordingly. That way, they will have fewer liability suits as well as fewer dead customers.

7

Learning from the Customer

Customers are a company's greatest resource. Organizations that understand and provide for their needs, wants, and desires will be rewarded. Those who don't are like leaves in the wind—unable to control their own destinies. Customers, however, are as hard to predict, anticipate, and understand as hyperactive three-year-old children on a diet of chocolate bars and sugar snacks.

Marketing surveys are traditionally conducted to determine whether a new product will sell. Most marketing departments, however, know that potential customers don't always say what they mean. If all marketing surveys were 100 percent reliable, Kaiser automobiles would still be in the show rooms.

Although surveys are helpful, companies should have supporting information on customer needs. Valuable sources of this feedback include:

- Unsolicited letters
- Quality assurance forms
- Warranty registration cards
- Distributor and retailer feedback
- Service agency reports

Many companies have the slogan: "If you like our product, tell others. If you don't, tell us." If you are lucky, some customers will oblige and send unsolicited letters. This is especially true when they receive outstanding service, are in a good mood, possess plenty of stationery, and have just read a book on social behavior.

In 1971, the Mobil Oil Company president received an unsolicited letter. It praised the company's marketing manager for checking on the activity of the writer's account. The marketing manager wanted to know if the customer's credit card had been stolen. It pleased the consumer to know that someone was looking out for him, even though he still had his card. The flurry of activity was caused by looking for a new job and traveling from California to Colorado after accepting one.

Despite feeling nice about special services, however, most customers need additional motivation before they write. In this instance, the additional motivation involved a gripe. The customer wanted to complain about a Mobil service station in Gallup, New Mexico; the station operator had talked the customer into buying new tires when the old ones were good for an additional 20,000 miles. In exchange for the feedback, Mobil gave the customer a complete refund and let him keep the tires.

Unsolicited customer feedback is valuable and should get broad distribution within the company; it should not be the personal property of the marketing department.

What should you do with customer comments? One large tea company sends duplicate copies of customer comments to the quality assurance department and the company's vice presidents. This procedure enables those involved to initiate corrective measures on their own. It also enables the quality assurance staff to identify trends and coordinate broadly based corrective action when necessary.

An Air Force installation went to the other extreme. They funnelled all customer comments through a single Air Force colonel. When the quality analysis organization tried to get copies, they were told they did not have "a need to know." Favorable comments were sent to the individuals or organizations that were complemented. Unfavorable comments were sent to action parties for response or corrective action. The quality analysis branch and other organization managers, however, never received copies. As a consequence, trends were never identified and potentially valuable feedback was lost.

Quality assurance forms, requesting customer feedback, are another source of valuable information; they are most popular in service industries such as fast-food chains and hotels. These forms, however, give management a biased and unreliable sample. If the forms are left for the cleanup crew to collect and process, favorable comments are sure to be processed. Comments criticizing friends may never get past the waste basket.

Quality assurance forms addressed to top management have another shortcoming. These forms seldom leave the establishment unless they are stamped, addressed, and easy to fill out. They also get lost in purses or coat pockets unless there's a mailbox handy. Few people go out of their way to provide customer feedback unless there is "something in it for them."

Warranty registration cards have an advantage over other requests for information. Customers are led to believe their warranty is not in effect until the company gets the cards back. In order to take advantage of this high return rate, some companies use their warranty registration cards to ask for information on how customers like the product.

Some warranty registration cards also encourage response by offering free subscriptions to user-oriented newsletters. This procedure is particularly popular with computer and computer software companies. The newsletters enable the company to keep customers informed of new products. They also provide the company with a media for feedback on problems and successes occurring in the field.

One shortcoming of data obtained from warranty registrations is that the cards are normally filled out before the customer has had time to evaluate the product.

When warranty cards are reviewed by the quality organization, they provide information that can be used for follow-up at a later date. Once the quality department knows who owns the product, they can send them questionnaires a year or two after the date of sale.

Inquiries, mailed to a sample of customers, help identify what they like about the product. Customer response is significantly improved when you offer some novelty gift for participating in the survey. A 40 percent participation has been observed when customers had a good reason to reply. Less than 1 percent participation has been reported when customers had no incentive to answer.

One company gets customers' feedback by offering them a new unit in exchange for one they bought a year before. The company explains that it is using the exchange as part of its quality program; this gives the company a chance to make detailed technological evaluations of the entire product after a year's use. If the offer is accepted, the company sends the customer a new unit with instructions for returning the old. The company also pays the postage.

Reports from distributors and retailers provide additional feedback. Unfortunately, they seldom get beyond the marketing department. Retailers are close to the customer. They are often aware of problems long before anyone else. When comments are routed through the quality organization, many small problems can be corrected before they become big.

Service and repair agencies can provide additional feedback on how products perform in the field. As an example, suppose your company makes toasters and your service shops report a large number of units have come in for new heating elements. This information would enable you to redesign the elements before marketing the next model.

Feedback from service and repair agencies has great value when the information is distributed throughout the company. It has little value when scanned and stored in someone's private file, even if that someone is the company president.

Units that are returned for repair provide additional information.

- Why did it fail?
- How did it fail?
- Is there indication of customer abuse?
- If customer abuse is common, should the design be changed to enable the product to survive that type of treatment without breaking down?

The consumer should never be your sole source of information on product weaknesses. When use or abuse by the customer identifies product weaknesses, however, accelerated aging tests should be developed to help you verify the mode of failure. Then you can change the design to eliminate the defect.

Another mode of feedback is to simulate the customer by taking a sample at the point of shipment. Then test it. Interference with shipment schedules can be minimized by having spare units ready to replace the samples.

The test sample should be small; the sample size should increase slowly as the production volume of the item increases. Sample size should also be influenced by the complexity of the product and the consequences of a defect. Sampling frequency should vary from daily to once a month depending on the quantities involved and the consequences of failure. If no defects are found in a year, these tests might be reduced or eliminated.

The sample selection should be random. This means that each unit within a container should have an equal chance of being selected. If there is more than one container, random samples should be taken from each, assuming they hold enough units to make this procedure reasonable. It would be impractical to sample each carton if they all contained a single unit.

When evaluating the sample:

1. Use a quality team to simulate the customer by asking the following questions.
   a. How do they react to printing on the package?
   b. Is the package too flimsy or too sturdy?
   c. Can the package come open during normal handling?
   d. Is it too hard to open?
   e. Are the instructions enclosed?
   f. Are the instructions clear, complete, and concise?
   g. Are the instructions in as many languages as needed?
   h. Is it too hard to get the product out of the package?
   i. What problems can be anticipated if the customer uses the product before reading the instructions?
   j. What would happen if the package was mistreated by delivery personnel? As an example, suppose you were shipping a dozen eggs. What would happen if the delivery crew tried to stuff the carton into a mail box? In this instance, the result is obvious. In some cases, however, the quality team will be unsure about some of the answers. They should then initiate testing and get the facts.
2. Try out the product as it is supposed to be used. This requires exposing the sample to all of the environmental conditions it will see during normal abuse.

3. Give the sample to the inspection department and have it tested for all important characteristics.
4. Run tests to determine what will happen if consumers abuse the product. Every conceivable form of abuse should be tried until the samples fail. This is the only way to find the weakest points of the product. When all samples have failed, the quality team should decide whether the design needs modifications. The product does not have to withstand unreasonable treatment. Abuse which can be anticipated, however, must be dealt with. This is especially true when hazards to life or limb are involved.

Testing provides a reasonable understanding of a product's ability to meet the consumer's needs. Companies must look at the right thing, in the right way, at the right place, at the right time, and with the right test equipment. They must also use the right test procedures.

The best proof of customer satisfaction, however, is a growing volume of sales accompanied by no complaints and no returns. Quality catastrophes seldom occur when a company takes time to determine and meet the customers' needs.

8

Key 2: Anticipating Defects

In order to anticipate defects, you have to know what they are. Otherwise, you'd be like the medieval knight who spent ten years looking for the Holy Grail without knowing it was a cup.

Technically, a defect is a characteristic that fails to conform to the specification. A defective is a part or product that contains one or more defects.

Specifications, on the other hand, are descriptions of dimensions, types of materials, capabilities, and other characteristics that define an acceptable product.

How do you decide what is acceptable? In the final analysis, customers determine whether a product is acceptable or not. Therefore, the specifications should reflect user needs. This means there must be careful exploration of customer reaction before new or revised products hit the market.

Preproduction samples should be evaluated by people within the company. Manufacturing should determine whether they can make the product. Quality control should determine whether they can inspect it. Finance should determine whether they have funds for production and inspection tooling. Field services should determine whether they can repair it in the field.

In some instances, relatives, friends, and company-sponsored test panels can help determine the attractiveness and durability of the preproduction samples.

47

Evaluations of preproduction samples should simulate the intended market. Testers should be told to give the product normal wear and tear; they shouldn't return it to the company in the same shape they received it. By using and abusing the product, they act like future customers—people who will expect the product to last forever, no matter what they do.

The time to find out how a product will withstand mistreatment is before it is on the market. This includes trying to find out what customers will do when they fail to read the instructions.

When people buy something, they want to try it out. They don't want to get bogged down reading a pile of vendor literature. This trait of human nature gave rise to the saying, "When all else fails, read the instructions."

Food and beverage companies frequently use test panels to determine whether a new product has customer appeal. Members of the panels usually come from churches, clubs, and other groups interested in raising money for their organizations.

Taste testers are normally paid for their opinions, even though they are not professional tasters and have received little or no training. Individuals are often limited to quarterly or semiannual assignments. Otherwise, they start anticipating what the companies are looking for rather than expressing individual preferences. Companies don't want biased evaluations.

Once the food or beverage has been accepted for production, a professional panel may set more objective standards or taste profiles for the product. Subsequent production can be checked against these profiles. When the product is ready, test marketing in localized areas can help determine whether the specifications are appropriate. Similar evaluations can be made on other products.

Companies have to anticipate problems. They can't afford to blame the consumer when their product is ruined by customer abuse. If it doesn't work satisfactorily after harsh treatment, the customer will usually blame the company and quit buying its products.

When you know what pleases and what displeases the customer, review the specifications. If changes are made, do additional testing to ensure that the product has been improved. In the end the design should result in a product that is safe, desirable, durable, and simple—and the cost must be competitive.

Specifications must cover the wide range of quality characteristics from critical to incidental. Outside dimensions that establish appearance and shape should have liberal tolerances. Parts that must mate with each other should have tighter controls. The dimensional specifications should not, however, go beyond ensuring that the parts can be assembled properly and fit without being too loose. The stricter the specification, the more expensive the manufacturing process.

Dimensional specifications are usually straightforward. The prime consideration is to establish tolerances that are realistic. Material specifications, however, are more difficult due to their complex nature. Materials should perform the intended function; they should also withstand customer abuse.

Usability, durability, and reliability specifications are usually the most difficult. They may be the most incomplete, also. Don't make your specifications so loose that suppliers can base their bids on price alone. Materials and parts must stand up in service.

Specifications should be written so there is no doubt about how inspection procedures are to be carried out. As an example, you may want to set rigid specifications on the artwork for your package. This usually includes providing the printer with samples defining maximum and minimum color intensities. If you fail to specify the light source in the inspection area, however, your specification is incomplete. Artwork may look different under a standard 5000 ° light than it does under a 7000 ° light.

Lopsided specifications should also be avoided. Designers who want to get as close to a dimension as possible, without exceeding it, often move the target value to one side of the specification.

Drawing dimensions of 1.065 inches +.005, -.001 are lopsided. In this case the designers want production to aim for 1.065. They plan, however, to reject parts that are below 1.064 or above 1.070. Since most dimensions tend to distribute themselves normally, production must consider the capability of its equipment and set its aim accordingly. If its machinery can only hold to a scatter of .003 inch, production must aim at the middle of the specification, which is 1.067.

If the designer really needs the bulk of the parts to center around 1.065 with a lower limit of 1.064, then the specification should read

1.065 ± .001. In order to meet this tighter specification, a more precise machine may be needed and the cost may increase.

Conformance to specifications is a much neglected aspect of quality. In many cases, it is also taken lightly. If a specification doesn't mean what it says, why have it?

The owner of a small plant once told a disgruntled customer, "So what if the parts are not inside your specification? They're only off by .001 inch." Actually, some of the parts were off by .010 inch; to the buyer, they were unusable.

The best approach to specifications is to make them as loose as possible to still be consistent with pleasing and protecting the customer. Then enforce them.

Look for critical defects like a small town cop running a speed trap. Don't relax, no matter how long it's been since you found a potential catastrophe.

Inspection of critical defects is like fire insurance. Would you cancel insurance just because your house hasn't burned down lately?

Major defects should be inspected with care using a random sample. No discrepancies should be accepted. A single unit that is out of specification by a cat's whisker may be the tip of an iceberg.

When a major discrepancy is found, sample more units to find how extensive the defects are. Then, based on this detailed information, decide on disposition of the material.

While minor quality characteristics should adhere to specifications, they rarely result in adverse customer reaction. To the extent feasible, check them when you inspect more important characteristics. Don't overlook them but don't nit-pick either. A dedicated nit-picker can shut down a smoothly running assembly line.

When inspecting major and critical characteristics be sure that the testing is done correctly and the test equipment is in good condition; it may malfunction due to misuse, wear, or drift. You should also ensure that the test procedures are correct.

Equipment used on critical characteristics should be checked daily against a standard and rechecked each time a defect is found. Standards must also be compared with master standards from time to time.

Make a routine of checking dimensional inspection equipment against Jo blocks. Check other measuring devices against their appropriate standards. Then record the results. Most measuring instruments, like temperature gages, pressure gages, and timers, become inaccurate after extended use. They should be calibrated on a regular schedule.

Your new electronic devices may be more accurate and more precise than your older mechanical equipment but don't take them for granted either. Check them thoroughly for both accuracy and precision before using them.

ATTRIBUTES VERSUS VARIABLES

There are two types of test equipment, those that measure variables and those that measure attributes.

Variables measurements give you numerical readings. They can show degrees, inches, pounds, pounds per square inch, or any other unit of measurement. When feasible, measure quality characteristics on a variables scale. If the part is good, you will know how good. If it is bad, you will know how bad.

Attribute measurements tell you whether something is good or bad but not how good or how bad. Attribute mesurements are usually made with go/no-go gages which may come in pairs. One go/no-go gage may show that a part is too small if it fits into the gage. Another gage may show that the same part is too large if it does not fit.

Go/no-go gages require less time to use than variables devices. While pieces can be inspected more quickly with go/no-go gages, they do not provide as much information as variables measuring equipment.

Although you have flexibility in selecting your test and measurement equipment, there is one area where compromise can be fatal. Don't use gages that are inaccurate or imprecise. It is embarrasing to reject a good incoming lot of material because your gages were out of calibration. It is even worse to accept a discrepant lot of your own product.

As the inspector said to Saint Peter, "I was sure the electricity was off. My voltmeter didn't give a reading and it didn't have a warning tag. I've got to admit, though, the meter didn't have a calibration tag either."

9

Quality Catastrophes

Quality catastrophes are like land mines. If you anticipate their presence and plan to avoid them, they seldom blow up. After one explodes, it's easy to see what you should have done—if you live through the blast.

Years ago, companies designed, built, and inspected their products for local consumption. Good units were sold. Bad units were reworked or discarded. Quality catastrophes were common, but most buyers accepted them as inevitable; they didn't complain too much.

Today, however, customers are becoming more sophisticated and choosy. For companies, economic survival demands that they stop making scrap. They must find out why catastrophes occur and use this knowledge to prevent recurrence. In addition, they must anticipate defectives and take corrective action before a discrepant product is sold. Steps for accomplishing these goals include:

1. Anticipate and avoid supplier difficulties.
2. Improve product designs.
3. Buy the right equipment.
4. Correct assembly problems.
5. Utilize inspection more effectively.
6. Improve preproduction testing.
7. Eliminate packaging and shipping catastrophes.

8. Conduct aging tests.
9. Anticipate product abuse.
10. Coordinate.

SUPPLIER DIFFICULTIES

Quality starts with your suppliers. If their quality is poor, your quality will be poor.

Have your purchasing, quality, and design departments visit your suppliers. They should work as a team, evaluating the vendor's ability and willingness to provide quality products.

Develop good lines of communication between your quality representative and the vendor's. The moment any critical defect shows up, contact the vendor to explain what was found. Make sure you and your vendors use the same equipment and procedures when inspecting parts and materials.

When buying production or inspection equipment, be particularly careful. Troubles may be encountered with any tooling or machinery, especially when it was developed for special purposes.

Be certain that the vendor knows:

• How you plan to use the machine
• What parts will be made on it
• What tolerances must be met

Have your operators make test runs on the equipment prior to delivery. During these tests, use your own parts and materials. This will help ensure that the machine will do the job without requiring special environments, procedures, or skills.

IMPROVING PRODUCT DESIGNS

If your product isn't designed correctly, there's nothing production can do to make it better. Typical design failures include:

1. Failing to include what the customers want
2. Specifying raw materials improperly

3. Setting tolerances that are too tight or too loose
4. Omitting specifications needed for the product's success
5. Setting tolerances that keep parts from mating properly
6. Failing to specify the right operation conditions
7. Setting requirements beyond the state of the art
8. Setting requirements that can't be met with available machinery

If you are going to avoid these problems, your marketing department must anticipate customer desires and coordinate these findings with design engineering. Then, the quality team should verify that the proposed design is acceptable. They should check safety, appearance, function, reliability, maintainability, repairability, and other appropriate variables.

Raw material selection is also critical. There is a tendency among designers to assume raw materials will be as the supplier represents them, will not be too variable, and will perform as the designer intends. Any one of these assumptions can result in a quality catastrophe.

The variability of raw materials should be determined during the vendor evaluation visits. If the vendor doesn't know what the variability is, get samples from different raw material lots. Then have your lab make its own determination.

Some raw materials will work well if the end product is used as the designer intended. The quality team, however, should ensure that the materials will stand up under adverse conditions resulting from customer abuse. Extreme conditions must be fully explored.

BUYING THE RIGHT EQUIPMENT

Manufacturing equipment can cause trouble in many ways. Old equipment can develop excessive play in the bearings. New equipment can have inadequate adjustments.

For example, one company designed a heat treating oven but failed to specify adequate time and temperature cycles. Once the oven was in service they found the optimum time-temperature cycle exceeded the oven's capability.

Every piece of equipment has its own working range; this range must be determined before the machinery is purchased. In addition, it must be designed with the correct features and adjustments. Otherwise, it will not produce good parts. This means the design engineer must run tests to determine the range of sizes, times, temperatures, pressures, voltages, speeds, and other variables that will be needed.

If you have several machines of the same kind, they will probably vary with respect to their accuracy and precision. If you have the choice of which machine to use for a specific job, be sure the one you choose can meet the product specifications.

A machine that can only produce parts to ± .010 inch. is not suitable for parts that must be machined to ± .005 inch. On the other hand, you might waste resources by putting a job with ± .010-inch tolerances on a machine that can be held to ± .001 inch.

An exact match of machine capability to product tolerances, however, is not always desired. Wear of bearings or other components is inevitable. With time, the machine will gradually begin to produce unacceptable parts. Some allowance should be made for machine wear.

CORRECTING ASSEMBLY PROBLEMS

There are two prime ways assemblies can fail. The first is when parts can be put together backwards. This can only happen when the mating areas are symmetrical. Situations requiring symmetry are rare. The designer, therefore, should include enough asymmetry to make incorrect assembly physically impossible.

Many microcomputers had this problem. You could insert mating connections upside down. In some instances, this burned out parts of the computer. Wiser users saw the problem and marked the tops of each connector. They also made markings to show where each plug belonged. Eventually, the microcomputer companies fixed their connectors with asymmetrical housing that kept users from mating them incorrectly. They also identified which plugs and sockets went together.

Another way assemblies can fail is for design dimensions or tolerances to add up improperly. Designers seldom call for parts bigger than the holes they go into, but this error can happen.

A third problem for designers is assurance that the tolerances are realistic. Unrealistic tolerancing was common in the aerospace industry of the 1960s. At that time many people believed that designers chose their tolerances by picking the smallest number they could think of and then dividing by two.

The designers used the logic that tight tolerances would shield them from blame in case the part failed. They were inconsistent, though. When parts were rejected due to production's inability to meet the tolerances, the designers accepted many of the rejects for further processing. The acceptance process, however, required the concurrence of engineers from the quality and production departments. In this way, the designers could share the blame rather than accept responsibility for their errors. They did, however, expect credit for their successes.

UTILIZING INSPECTION MORE EFFECTIVELY

Many people think inspection is the simple act of comparing a part or assembly with a specification to determine if it is good or bad. Unfortunately, it is not that simple. Inspection can be performed with an infinite variety of gages, some of which are quite complex. The greater the precision required, the more complex the gage is apt to be.

One step forward was taken when the micrometer was invented. The ordinary hand-held micrometer is supposed to measure to the nearest .001 inch and is widely used for this purpose. Its greatest weaknesses are:

- Micrometers get out of adjustment; they should be checked frequently.
- Their threads wear.
- Their accuracy depends on the touch of the inspector, although this is partially compensated for by ratchet knobs.
- They are not able to accurately measure to .0001 inch, even though some models have verniers that give readings in this range.

The accuracy and the precision of measuring equipment is one factor influencing inspection. Other factors include:

1. **Boredom:** The maximum number of consecutive pieces that can be inspected effectively is about 12. The exact number is influenced by the nature of the inspection procedure and the item being inspected.
2. **Distraction:** Inspectors might be distracted by a miniskirt swishing by or by a hunk from Muscle Beach passing through on a tour.
3. **Fatigue:** Large volumes of work induce fatigue. So does excessive overtime or disruptive shift schedules.
4. **Poor lighting:** Although most complaints come from dimly lit areas, the glare of excessive light can also hamper good work.
5. **Inadequate instruction:** You shouldn't blame inspectors for doing a job wrong if they've never been told how to do it right.
6. **Poor equipment:** It's hard to do a good job with gages that are worn or out of adjustment.
7. **Poor eyesight:** This problem is particularly acute when the product is small, intricate, or complex.

There are many automatic, sophisticated inspection systems available today. This is fine when the characteristic is critical, 100 percent inspection is required, and volumes are large. Be sure, however, the equipment is calibrated, works correctly, and gives the right results.

For additional protection, feed known defectives into the system periodically. If the equipment doesn't catch the defectives, shut the system down and have it recalibrated. One of the defectives should be out of specification on the high side and one on the low.

When inspecting critical items without having automatic equipment, use backup or multiple inspection techniques. As an example, suppose the inspector is checking a circuit board for the presence of a critical resistor. Just looking isn't enough. After monitoring a large number of boards, the inspector may "see" the part even if it isn't there.

In addition to looking, have the inspector touch the resistor with a stylus. Then have the inspector place a template over the board, highlighting the resistor. With three independent ways of checking, errors should be reduced to the vanishing point.

IMPROVING PREPRODUCTION TESTING

Preproduction testing may involve physical stresses, chemical action, temperature variations, humidity changes, electrical loads, or performance requirements. One common mistake is to test the product under ideal conditions.

Unfortunately, some reasonably common situations may be overlooked. For example, a weak part of an assembly might make a convenient handle for picking up the unit. The designer should anticipate this and beef up the weak part. An alternative is to reposition the part so customers can't get their hands on it.

Overloads are another problem. For example, ladders must not only carry people, they must also carry the heavy equipment taken with them when climbing. Preproduction testing should anticipate this problem and determine whether the design can survive it.

Chairs are made to be sat on. People, however, will use them for ladders. This must also be considered in preproduction testing.

Ironing boards are made for ironing, but people load them with bags of groceries and other heavy items. They should be tested accordingly.

Environmental testing is also necessary. Devastating chemical reactions occur along the seacoast where salt spray turns into dilute hydrochloric acid. Acid attacks metallic materials, like automobile bodies. The result is rust. Salt, used for melting highway snow, is even more corrosive.

At least one helicopter hub fell prey to a similar problem. It was designed to handle normal operations but it failed in flight. A crash investigation uncovered the fact that the hub was not resistant to corrosion. Dew, carbon dioxide, sulfur, and other atmospheric contaminates combined to attack the metal. The attack made it weak, causing it to break down.

On the other hand, electrical loads are not much of a problem in modern homes and factories. Few people know how to bypass the overload devices that protect the circuits. Unfortunately, some electrical design engineers don't get off as easy.

For example, the starter subassembly on an industrial air-conditioning compressor had design problems. Soon after the first units were installed, three subassemblies failed. Factory trouble-shooters found that one specific condenser burned out in each subassembly. The electrical surge at start-up was great enough at times to wreck the condenser. Preproduction testing should have caught this problem.

Makers of home appliances face special hazards. Electrical shocks occur when least expected. In addition, children are inquisitive and try to take things apart. This has lead to a flood of "child-proof" caps for electrical outlets. More safequards are needed on household devices to protect both children and adults.

As an example, electric cords hang down to the floor where children can pull them. Heavy appliances, pulled off the counter, can do a lot of damage to youngsters. These appliances should have pull-apart plugs close to the base of the device. If properly designed, the plugs would come apart so easily that the device would not move when the cord was pulled. Then, only the soft rubber plug would hit the child. As far as we know, no manufacturer has ever sold such a device.

Food processors present another problem. A recent TV program for consumers described several makes of these appliances. The high-speed blades in the machines can cut fingers, hands, and arms. To avoid accidents, all of the processors were equipped with safety devices that would shut off the motor if the cover was removed.

One model, however, permitted the cover to be lifted far enough that you could reach the blades with your fingers before it shut off. Needless to say, this model was not the safest unit in the show. The average user might not learn of this deficiency until too late.

Performance requirements are usually spelled out by the design engineer. The quality team should make sure they are adequate.

A common oversight of design engineers is to fail to anticipate customer abuse. They should also look for environmental conditions that can occur. For example, most faucets are designed for use with clean, clear water. Iron pipes in old homes, however, are often full of rust. Even copper pipes in new homes can pick up large amounts of dirt before being installed. While plumbers are supposed to flush lines thoroughly, they don't always do so.

One company put out a line of valves which were unusually easy to damage when exposed to rust, scale, sand, and dirt. These valves cost them a fortune in warranty claims and lost sales.

Many of these problems can be predicted by running life tests using devices that repeatedly turn the units off and on. The product is run through innumerable cycles until it fails; counters are used to record when they quit working. Using this data, statistical engineers can identify failure patterns and predict which parts will fail first.

These life tests are good. The testing, however, should expose the units to the most severe environmental conditions they will experience in the real-life world.

During life testing, units are cycled to destruction. These tests reveal the weak points of the product and help identify where corrective action is required. All manufacturers should know the conditions under which their products will fail.

ELIMINATING PACKAGING AND SHIPPING CATASTROPHES

In order to determine what stresses and shocks occur during shipment, sensors can be installed in test packages. With the information obtained from these sensors, companies can develop programs to ensure their products will survive hazards equal to or greater than those encountered during the tests. Companies need to be sure that their products reach the customer in as good condition as they were when they left the plant. When running shipping tests, take into account the method of transportation, the durability of the products, and the distance they will travel.

Big bold signs that say "this end up," "do not stack," "Open here," and "fragile" may prevent some damage, but don't depend on them. It is safer to design your package so it can ride upside down, be stacked ten high and be opened anywhere. If your product is fragile, provide extra protection.

You also need to know what will happen to the product if it is exposed to extreme heat or cold during shipment. Trains and trucks travel through mountains and deserts at all times of the year. Some products may have to be shipped in refrigerated or heated transports.

Wrecks also occur. It is generally not economical to provide protection that will survive the severest wrecks. Economics or practical considerations should be considered.

The shipment of ballistic missiles, however, is one case where packaging is critical. What would happen to the neighborhood if a truck carrying a poorly packaged missile were in a head-on collision with another semi?

Flight recorders on commercial aircraft are also expected to survive the most horrible wrecks. Here the objective is to prevent a repetition of the accident. The flight recorders are designed to provide information on what caused the accident. Then corrective action can be taken to prevent recurrence.

CONDUCTING AGING TESTS

There was a time when batteries, rubber items, and other complex organic products didn't last long, even when they were sitting on a shelf. Although technological progress has extended the lives of these products, shelf-life must still be considered.

One protection against shelf-life catastrophes is to date products. This is particularly true with medicines and food items. They may sit on customer shelves for years and eventually become ineffective, foul, or dangerous.

When companies want to extend the shelf life of products, they often refrigerate them until sold. They also recommend that customers refrigerate them until consumed.

One of the top ten brewing companies stores and ships its products under constant refrigeration. Without refrigeration its shelf life is greatly reduced since the beer is not pasteurized. No one wants to drink a skunky beer.

The severity of shelf exposure varies dramatically. Time, temperature, sunlight, dust, and humidity must all be taken into account.

Stock that hasn't reached the store shelves may degrade in the warehouse. Consider this possibility when packaging your product.

The time between manufacture and sale is often evaluated by marketing and sales. They enclose a warranty card and ask customers to return it promptly. The date of production is either coded on the card or determined from the product's serial number. Based on this information, the company can determine and make allowance for the time it takes to move products from the production line to the customer. Consumers may accept a product going bad after sitting on their shelf for years. They rebel, however, if the product passed its useful life before they bought it.

ANTICIPATING PRODUCT ABUSE

Most designers fail to make allowance for customer abuse. It is, however, one of the most widespread causes of product failure. Many people either do not read instructions or forget them. Other customers expose products to devastating stresses that are hard to anticipate.

People who need a hammer and do not have one available will substitute almost anything in sight. Production operators have been known to use micrometers and other precision measuring devices to pound on things. Similar misuse occurs when something is needed for prying. People who abuse products seldom blame you if they damage the product. They will blame you, however, if they loose a finger, hand, or eye.

You cannot economically design your product to stand up under all the punishment customers might inflict. You do have the responsibility, however, to anticipate as much of it as you can. Then design your product to bear up under those stresses if it is economically possible do so.

COORDINATION

One person can anticipate certain hazards; another will anticipate others. Few people see exactly the same problems in the same way.

Group coordination by the quality team uses this concept to turn the spotlight on a wide variety of hazards. In addition, they use the catalytic action that occurs with the interchange of ideas. Without group action, many hazards would be overlooked.

Coordination, anticipation, and planning can help your company prevent quality catastrophes to the extent that:

1. Scrap and rework will be drastically reduced.
2. Customer complaints and returns will be decimated.
3. Recalls will no longer be necessary.
4. Liability suits will become a thing of the past.

Virtually any quality catastrophe can be prevented if it is anticipated. Teamwork is the key to anticipation.

10

Management Controllable Defects

One of the most quoted passages in the gospel according to St Luke is, "Physician, heal thyself." Does this apply to management?

It is easy to blame poor quality on the younger generation, the drug scene, disinterested workers, foreign competition, tough times, and hard luck. At least part of the problem, however, lies with top management. It controls the philosophy and the resources of the company. Good quality starts at the top.

If management wants excellence, it should provide:

1. Financial stability
2. Good raw materials
3. Adequate production equipment
4. Acceptable environmental conditions
5. Necessary training
6. Understanding

FINANCIAL STABILITY

Financial stability helps companies stay in business. To a large extent it is achieved through management's wise use of corporate resources.

Does top management always pursue financial stability? No, not always. As an example, some executives gut company finances in order to ward off corporate takeovers, hoping to save their own jobs in the process.

At one time, the payment of greenmail was a common ploy. Greenmail is a corporate type of blackmail. It occurs when companies pay excessive prices to get stock back from raiders threatening to take over the company. In most instances, greenmail hurts the company's employees and investors. Excessive payments weaken the corporate balance sheet.

Although greenmail may be restricted by recent legislation, similar corporate maneuvers are still being practiced. As an example, corporate executives may vote for "golden parachutes," where they get excessive severance pay should a takeover succeed. Comparable treatment for lower level employees is almost unheard of.

Another ploy is corporate restructuring. Some restructuring puts the company so deeply in debt that no corporate raider would want it. Again, financial stability is forfeited and the company spends years trying to recover. In some cases it succeeds.

Many corporate takeover defenses hurt the company's employees by draining off funds that could be used to buy tools and equipment they need on the job. They may also lose jobs if the company retrenches because of squandered resources.

GOOD RAW MATERIALS

During the early days of the space program, a common saying was, "Would you like to go to the moon in a rocket made by the lowest bidder?" In the space program, this was a joke. The critical nature of space missions caused the government to ensure that aerospace contractors were qualified to do the work.

In commercial enterprises, the saying takes a bitter twist. More than one purchasing agent has been commended for finding cheaper parts and raw materials. Often, however, the savings vanish when scrap, rework, downtime, and parts shortages are added to the cost. What good are cheap vendors when their products cause excessive production problems?

Management can improve raw material quality by ensuring that the purchasing staff consults the quality department and the quality team before selecting a vendor.

The quality department should identify the quality of goods the vendors have been supplying.

The quality team should identify special requirements for the purchase order. As an example, corrosion resistance, shelf life, stable plasticizers, or nontoxic ingredients may be logical parts of the requirement. Many of these special needs should be identified by pre-production and life tests monitored by the quality team. As the saying goes, "You can't make a new silk purse out of an old football."

PRODUCTION EQUIPMENT

Management's responsibilities include ensuring that production and inspection personnel have equipment for doing the job right. Inadequate equipment causes as many problems as shaky raw materials. You can't meet ± .001 tolerances with machines that are only good for ± .010.

Some highly skilled operators get more out of old, rickety machinery than others. Even then, the price is high. Pushing equipment to its ultimate capability requires more time and skill than routine production. When management skimps on equipment, they usually pay more for scrap, rework, downtime, and wages.

Management shouldn't buy cheap inspection equipment either. How can you meet customers' demands without accurately measuring what you made? A common rule of thumb is, "Inspection equipment should have ten times the precision of the specification requirement." This rule implies that inspection gaging must be accurate to ± .0001 if the specification has dimension requirements of ± .001.

The above rules apply whether the parts are measured by operators or inspectors.

ENVIRONMENTAL CONDITIONS

During the 1950s, one small West Coast petroleum refinery had the reputation of being a death trap. Their fatality rate, however, wasn't unusually high. The operators were well aware of the risks and protected themselves. Today plants like that would be shut down by the Occupational Safety and Health Administration (OSHA) the moment a worker complained.

Management must ensure that the working environment is safe. In many cases, they must also ensure comfort. Companies that don't provide adequate lighting, noise control, and temperature control are apt to have disgruntled employees.

What are disgruntled employees? They are workers who are more interested in backing up to the pay window than helping the company stay in business. "Backing up to the pay window" once described workers who were ashamed of their performance. They couldn't look the paymaster in the face.

During World War II, many weather station managers were plagued by problems with the work environment. Air conditioning was uncommon and enlisted men frequently plotted weather maps when the temperatures and humidities were both around 95. Despite the liberal use of sweat bands and blotters, many weather maps were splattered with sweat. The quality was poor. The maps were hard to read.

Today, similar conditions are found in many "sweat shops." Does temperature still affect quality? Are people as conscious of quality when temperatures are miserable and ventilation is poor? Management must evaluate the tradeoff between the cost of air conditioning facilities and the slowdown of uncomfortable workers.

TRAINING

Many workers learn their assignments through on-the-job training. Mary trains Jim. Jim trains Sue. Sue trains Mark. Soon the job evolves and becomes unrecognizable by those who set it up. That is why management started using procedures and standards. You need a standard to minimize unintentional changes in the process.

There are times when on-the-job training isn't enough. Safety training is a good example. In the 1950s one West Coast refinery manager was demoted to a staff job because the safety training in his plant was inadequate. The plant had a leak of poisonous gas; by the end of the crisis, one operator became a human vegetable.

The workers in the area had adequate equipment but weren't trained to use it instinctively. When refinery accidents happen, you have to do the right thing without thinking. Doing the wrong thing can kill you. Moving too slowly can also be fatal.

Shortly after the accident, all personnel in the operations areas were required to go through weekly safety drills. Few drills were missed. Everyone knew that the lives at stake might be their own.

Management can also provide special training. Classes in statistical process control are a good example. Training of this type is most valuable when the operators are encouraged to use their new found knowledge on the job. As an example, when operators plot their own control charts, they can spot trends as they take place. In many instances, production personnel can relate process changes to something they observed. This insight into the cause of problems speeds corrective action.

Special training classes for managers tend to emphasize management. In many cases it would be more appropriate for them to study statistical process control or computer applications. Managers who know nothing about statistical process control have a hard time understanding recommendations made by their quality engineers. They, however, have to make the critical decisions about which type of corrective action is required when problems arise.

Managers who do not understand the language or the tools of statistical process control are likely to make poor decisions in that area.

Statistical process control specialists are not the only people to use special tools and foreign-sounding languages. Computer analysts work under similar handicaps. Managers can't make good decisions on data base quality control if they don't understand the language. Unfortunately, few managers realize that data base quality is a potential problem. Fewer know what to do about it.

Although training is associated with classrooms and lectures, quality circles can also improve employee skills. The trainers in quality circles, called facilitators, do some teaching. The greatest value of the circles, however, comes when they apply statistical techniques to plant problems.

Quality circles and their various clones have received mixed reviews from management. When a good trainer is available and top management supports the program, circles are usually successful. When management supports waivers, however, the programs fall flat. This gives management the excuse to say, "I told you so." In most cases, however, the problem is with management rather than the quality circle concept or the participants.

One of the least heralded benefits of quality circles is the training provided. Members are taught to identify problems, determine alternate solutions, make presentations to management, and implement findings. This training gives the companies a reserve of experience in supervision and management skills that can be drawn on as needed.

UNDERSTANDING THE TROOPS

Owners of small businesses usually know their employees by name; they spend hours on the shop floor every day. As businesses grow, layers of management increase.

Eventually, things change. Managers in large companies are considered diligent when they talk to their own staff every day.

Some managers keep in touch with their personnel by touring the plant before going to their office. This way, they get down to the grass roots before distractions of the day absorb their time. Managers who don't have time to listen don't have time to learn what's going on. It is difficult to anticipate, identify, and help control defects without having good lines of communication.

Managers should also minimize intimidation. Excellent ideas are lost when the originators are afraid of the boss. In one case, an aerospace engineer was so intimidated by his boss that he couldn't give an intelligent briefing. Realizing he had a problem, the engineer joined Toastmasters International and learned how to give better presentations—planned and extemporaneous.

The boss benefited from the engineer's additional training. The engineer, however, was stuck with an intimidating manager who had no intention of changing. The company suffered from the manager's love of power. In this case, the boss was more interested in exploiting his power base than understanding his subordinates.

Managers show understanding when they develop equitable systems for recognizing workers. One company uses a team approach for developing new products. All employees are encouraged to suggest new items; this includes janitors as well as vice presidents.

The one who has the idea tries to sell it to fellow workers in other departments like marketing, quality, production, and design. If the idea is good, volunteers from these departments help prepare a proposal for top management.

If the idea looks commercial and passes the first screening, a formal product development team is organized. Participants include all levels of the company. The criteria for selecting members is that they must have special skills and must be willing to work on the project on their own time.

The teams evaluate the product for customer interest, cost of manufacture, availability of raw materials, and other characteristics they deem necessary. Then they present their findings to top management.

Employees are not paid for their work on the teams. They do, however, receive bonuses if the product is successful. To date, at least one of the company's five top selling items was conceived in this manner. Many others are being marketed.

This same company has a newsletter that publicizes the accomplishments of employees on the job. In addition, it recognizes achievements at weekly departmental meetings.

Is the approach successful? The company thinks so and its 15 percent per year growth rate implies they are right.

11

Worker Controllable Defects

If you read enough books on participative management, you might conclude that workers are perfect and all defects are caused by supervision. Don't believe it.

Workers, supervisors, and managers are human. They all have strengths and weaknesses.

Variables influencing worker controllable defects include:

1. Health
2. Freedom
3. Paternalism
4. Individualism
5. Training
6. Worker attitude
7. Management attitude

HEALTH

Health has a major impact on performance. When workers are under the weather, they don't perform to their full potential.

Many health problems require the services of a doctor. Others, including good sleeping habits, adequate exercise, and reasonable diet, can be controlled by the worker.

One quality engineer kept falling asleep on the job. His supervisor suggested that he see a doctor, which he did. The doctor gave him medication to improve his metabolism. Although medication didn't completely solve the problem, it brought the engineer's performance to an acceptable level.

A technician had a more serious problem. His performance varied from being the best in the shop to being the worst. His problem was diagnosed as a chemical imbalance. Medical science helped a little but the wide swings in performance continued. Fortunately, the individual did so well when he was fully functional that his bad periods could be tolerated.

Sleep is fundamental. You can't get ahead if you aren't awake when promotions are being handed out.

Physical exercise is also important—much more important than most people realize. Exercise, performed regularly and in moderation, releases tension and reduces the probability of heart attacks. Some companies encourage exercise by having jogging breaks during the lunch hour or having gymnasiums for use after work. In both instances, however, shower facilities are recommended.

Drug and alcohol abuse are extreme cases of poor diet. They contribute to many health problems and they are worker controllable. When drugs and alcohol get out of hand, however, workers need help. Many companies provide rehabilitation facilities for those who get hooked. In some instances, the patients include supervisors, managers, and executives as well as clerks and operators.

Some health problems can be overcome by sheer determination. During World War II, one of the best technicians in a weather station at Gulfport, Mississippi had a glass eye. He didn't let it bother him, even though the job involved extensive detail work. When he got tired, he just took out the eye and polished it.

If you want your performance level to be high, keep healthy. If you want your subordinates to do well, encourage them to keep healthy too.

FREEDOM

Freedom is both precious and fragile. Too little freedom can cause workers to become uncooperative and rebellious. Too much freedom can cause them to make unauthorized changes to your product.

Procedures are one tool for limiting the misuse of freedom. They were developed to record the "right" way of doing things. When written and applied properly, procedures help companies produce consistent, high-quality products. When too restrictive, however, they cause problems.

Excessively tight procedures can become a challenge to the independent worker. One grind station operator in a West Coast propellant plant was particularly jealous of his independence. When procedures were written for his grind station, he claimed they were a handicap, not a help. Then he boasted that he could make good oxidizer with no instructions from anyone. He also boasted that he could make unacceptable oxidizer without deviating from the tightest specification anyone could develop. There was enough truth in his claim that the engineers didn't ask him to prove it.

Freedom is sometimes restricted when organizations want to give high-skill jobs to unskilled people. Civil service personnel procedures are a good example. Civil service rules are so detailed that even competent managers have difficulty filling openings with the best candidates.

The basis for tight civil service regulations is the government's desire to ensure that everyone is treated fairly. They want to eliminate biases favoring any color, race, sex, or creed. As Publilius Syrus said during the first century B.C., "It is a very hard undertaking to seek to please everybody." Many bureaucrats and politicians could profit from that wisdom.

The government's objectives are commendable. Their methods of accomplishing them, however, are self-defeating. How would you like to have a computer make all your decisions for you?

PATERNALISM

Paternalism refers to the "Father Knows Best" style of management. Excessive paternalism, however, hurts quality and productivity.

As an example, one Colorado company used to be known as a benevolent dictatorship. Anyone who followed the company line and didn't make waves was judged with leniency. Those showing too much individualism and those questioning the benevolent dictators became expendable. As a consequence, innovation was curtailed and the company lost ground to competition.

Governmental paternalism can have the same impact on quality and productivity. In their attempt to be fair to all the workers, many governmental agencies make it difficult to reward the good performers and discipline the incompetent. As a consequence, the best performers tend to leave the government; the worst retire on the job.

Thirty years ago, paternalism may have been helpful. It was nice to know "Big Brother" cared. In this age of participative management, however, paternalism tends to hurt productivity, quality, and creativity.

INDIVIDUALISM

Many managers assume that all people have their own goals, ambitions, and motivations. This is not always so. The quest for quality requires that each worker be treated as a unique individual.

As an example, a supervisor in an aerospace plant wanted to improve the versatility of his group by cross-training technicians. The jobs were switched, training was completed, and everything went smoothly for a week. Then, one of the technicians became moody for no apparent reason. On investigation, the supervisor found that the technician simply liked her old job and didn't want the new one. Job enrichment was not one of her top priorities.

The supervisor eventually restored tranquility by canceling the cross training. It hadn't occurred to him that different workers might have different goals. Some workers prefer a comfortable routine. Others demand a challenge. The supervisor would have been wise to determine which type of person he was working with before making the original change.

Converting expert technicians and scientists into mediocre supervisors can also cause quality problems.

A competent research chemist was promoted to laboratory supervisor because of his string of accomplishments. Soon, the productivity of the lab started slipping. The chemist's heart remained in the research area, despite his job assignment. "Demotion," back to the old job, was traumatic for both the chemist and the supervisor, even though the chemist didn't take a pay cut. In the end, however, the change paid off. Within two years, the research chemist had completed a number of sophisticated projects netting the company millions of dollars.

For maximum quality and productivity, you have to know your people. They are not all alike.

TRAINING

Some managers think that training is the answer to all plant problems. Too little or inappropriate training, however, causes difficulties.

Too little training results in people being expected to do jobs they haven't been properly taught. That is risky.

Inappropriate training can be just as bad. As an example, many nonsupervisory people request courses in management. If they are not in line for supervisory positions, however, management training often causes more problems than it solves.

If unable to apply what they learn, many people taking management courses critique their supervisors and managers. In the end, the students and their managers may become frustrated and unhappy. Few bosses take kindly to being told how to manage, and few students understand their boss' problems.

The bosses are not always wrong. Supervision and management is so complex that textbook solutions may not apply in many cases.

Special courses in technical writing create similar problems. Each organization has its own guidelines for technical reports. If the instructors and the students come from different organizations, problems will arise. Techniques, formats, and procedures taught in the classroom may not apply to the students' organizations. When that is the case, the students get confused and frustrated.

The best courses in technical writing are usually given by instructors from the student's own organization.

One way of getting more out of your training dollar is to have all students teach others once they complete the course. Advantages are:

1. Students have to digest what was taught before starting to teach. In many cases, they learn more as teachers than they did as students.
2. When they become teachers, students usually learn to apply some of the principles they were taught in class.
3. When teaching chores are spread around, more students have a chance to learn.

Some companies use conventions like ASQC's regional and national conferences to help their technical people keep up to date. Conventions, however, are expensive. In order to use them effectively, many companies require that their attendees:

• Identify the sessions they plan to attend
• Prepare trip reports, explaining what they learned
• Give classes to help those who didn't go
• Place conference transactions in the library
• Understand that transactions belong to the company if the company pays the fees

Training is an effective tool for improving productivity and quality. The above illustrations should help you make your training more effective and profitable.

WORKER ATTITUDE

We are slowly evolving into a society where people think that the world owes them a living. Illustrations of this attitude include:

One bright and effective technician quit her job when she found she could make more money on welfare. The government "owed" her the welfare check and she didn't want to be penalized for working.

An aerospace technician refused to participate in a problem-solving session. His explanation was that the engineers were being paid to solve problems. If they wanted his help, they should pay him engineering wages.

A promising technician applied for a transfer because she saw too much loafing in her department. She was a hard worker and expected everyone else to work just as hard.

A government clerk ranted and raved at the lead technician, confident there would be no reprimanded. When the supervisor initiated a reprimand, the clerk insisted that a union steward be present. Fortunately for the supervisor, the clerk started shouting at the union representative during the meeting. As a consequence, the reprimand was upheld.

How can worker attitude be improved?

An objective system for rewarding excellence and correcting incompetence helps. So does a good management attitude.

MANAGEMENT ATTITUDE

Workers are quick to see what management does; they are skeptical when they listen to what management says. As a consequence, when top management assigns a castoff to take over the quality function, the workers quickly realize the company isn't interested in quality. In cases like this, the attitude problem belongs to the management, not the subordinates.

The head of the quality department should be a quality professional, preferably a professional engineer (quality) or a certified quality engineer.

Workers also observe the attitude of management when evaluating subordinates. Is production volume the only thing that counts when it comes to appraisals, bonuses, and promotions? If so, then quality is getting lip service and nothing else. Quality suffers.

Management attitude also hurts morale when bosses:

- Fail to praise workers when they do a good job
- Praise them when they do a poor job
- Evaluate them subjectively

Textbooks are full of reasons why you should praise workers when they do a good job. Few books consider the impact of praising them when they do a poor job.

A laboratory supervisor in a West Coast petroleum refinery was always looking for opportunities to praise people. The chemists appreciated the attention and morale was high. Then one of the chemists messed up an assignment. Instead of discussing what could have been done to improve the situation, the supervisor continued to tell the chemist what a good job he was doing. From then on, the supervisor's praises had little effect on the chemist. He wanted to be praised when he did a good job but he didn't want praise when he dropped the ball.

Some supervisors give the same problem a different twist. Their evaluations are subjective or random. Awards they hand out are based on friendship, seniority, time since the last award, or the name that comes out of a hat.

Awards and pay raises are excellent tools for showing appreciation for a job well done; few managers use them efficiently.

In many cases, everyone in the organization gets the same percentage raise, whether their performance is good or bad. This approach results in "cost of living" raises. When performance isn't taken into account, however, these raises are considered a right rather than a reward. Fortunately, merit raises are becoming a more common method of recognizing exceptional performance.

Do you want raises and awards to mean something? Then give them for merit whenever possible. Automatic raises encourage mediocrity, not excellence.

12

Key 3: Quick Detection of Defects

Key 3 is the quick detection of defects. The sooner you find and correct these problems the less damage they do.

As Benjamin Franklin (1706-1790) wrote in **Poor Richard's Almanac,** "A little neglect may breed great mischief . . . .for want of a nail the shoe was lost; for want of a shoe the horse was lost; and for want of a horse the rider was lost."

Similar sayings have been attributed to George Herbert (1593-1633) and others. Some of these authors go as far as blaming the missing nail for the loss of a kingdom. These days, however, loss of a spark plug or rivet might be more appropriate.

Why would the horseshoe be shy one nail? Did the vendor shortchange the blacksmith? If so, was this the first time the vendor proved unreliable?

Was a shipment of nails rejected due to poor quality? Was a discrepant shipment tied up by the material review board when the shortage occurred?

Was the vendor the sole source of nails? Were other vendors available?

Was the nail vendor selected because of price or a cozy relationship with the blacksmith?

Was the blacksmith at fault? Had he overlooked the missing nail while rushing to get the job done?

Should the rider have performed receiving inspection on the work before riding off to battle?

Was the nail supplier hit with a big liability suit?
Was the vendor beheaded before the victor took over the town?
The problem could have been solved at many points in the chain, but it wasn't. A little neglect caused a catastrophe.

Detection of defects can occur:

1. During the design phase
2. At the vendor facility
3. During receiving inspection
4. During and after production
5. While being used by the customer

Related material on defects is included in Chapters 13 and 14.

DESIGN PHASE DEFECTS

The best time to catch problems is during the design phase. If a product is poorly designed, the best vendor, production, and inspection personnel can't do much to improve it. Quality teams and preproduction tests, however, can help you detect design problems before the job is assigned to the manufacturing department.

VENDOR DEFECTS

The second best time to catch a defect is before it leaves the supplier. At this time you don't have to sort out the bad material, rework the repairable parts, and dispose of the unrepairable. You don't have to prove who was at fault. You don't have to negotiate with the vendor on disposition, returns, or price breaks. The problem doesn't belong to you; it belongs to the supplier.

Many companies send representatives to vendors' factories whenever problems are anticipated or new products are introduced. Then, if the vendors can't understand the specifications, the company representative interprets them. It is best to clarify requirements when questions arise. If vendors make their own interpretations, poor guesses may not be corrected until several discrepant shipments have been delivered.

Anyone who can create fool-proof drawings and specifications can walk on water without knowing where the stones are hidden.

During normal operations, it may not be necessary to have personnel at the vendor site. When problems arise, however, send your people right away. Make sure the supplier's problems remain supplier problems, not yours.

Ship-to-stock (STS) and just-in-time (JIT) supplier control systems help ensure vendor quality. They require thorough screening of suppliers before contracts are awarded.

These programs are time consuming and expensive. They also reduce the number of qualified vendors. In the end, however, you get the best, not the cheapest.

The best suppliers are those who give the greatest value per dollar—after processing, warranty and liability costs are counted.

RECEIVING INSPECTION

If you are not using JIT or STS vendor control systems, receiving inspection is the next place for screening vendor problems. Here, good sampling, testing, and vendor liaison are critical.

When phasing in a new product, be sure that your vendor and your receiving inspection personnel run the same tests using the same procedures. They should also use the same equipment.

Laboratory and inspection personnel from the two companies should work out the details. It is better to discuss laboratory and testing procedures during the preproduction phase than argue about who was right after shipments have been rejected.

If it is not practical for the vendor and buyer to use identical equipment, run round-robin tests. Round robins involve having both organizations run tests on identical parts or products without comparing results until the tests are complete. Then look for inconsistencies between the buyer's and seller's results.

Problems with accuracy are found when one organization gets answers that are significantly different from the other's. Problems with precision are found when the results of one organization are more variable than the other's. These differences should be resolved as early in the procurement process as possible. The quicker you find a problem, the less it costs.

PRODUCTION DEFECTS

When vendors supply defect-free material, production problems are reduced but not necessarily eliminated. There are times when the drawings and specifications are not able to screen out materials that cause problems in your plant. This is particularly true when you're pushing the state of the art.

Problems at a large can plant during the 1970s are a good example. Some aluminum coil deliveries passed all receiving inspection tests but the metal wouldn't form into good cans. This caused problems between the vendor and the company.

Some operators wanted to reject the raw material as soon as they started having operating problems. In many cases, however, the operators were at fault; those with more skill and tenacity had little trouble using the material. In other cases, the metal wouldn't run smoothly regardless of the machine or operator involved.

Plant metallurgists tested the questionable material. If it checked out, they would run it on a different machine. If it still caused excessive scrap, they would work with the vendor to pin down the problem. By sharing information with the supplier, the company metallurgists eventually developed tests that helped screen the questionable aluminum. In the process, they advanced the state of the art. Both the company and the vendors benefited.

The production line should not be used for screening questionable incoming materials if there are other alternatives. The ultimate price in labor, scrap, rework, and vendor negotiations is too high. Shutting down operations due to the rejection of unacceptable raw materials is also expensive.

If the vendors slip in junk, receiving inspection should catch it. If receiving inspection fails to screen unacceptable material, the production line should catch it. If the production line produces an inferior product, inspection should find it. If final inspection lets low-quality products slip through, the customer will catch it.

The user provides the final screen for finding defects. This includes discrepant raw materials, parts, and assemblies.

Problems with having the customers find your mistakes include:

- Few customers suffer in silence. They may not complain to you but they will tell their friends. Then they and their friends will quit using your product.
- Defective products, in the hands of customers, may lead to a rash of liability suits.
- Recalling a product can be much more expensive than correcting the problem before it goes to the customer.

SUMMARY

The quicker you detect and correct problems, the less they will cost. A defect that is identified during preproduction testing can be an annoyance. A defect that is identified in the product liability courts can be a disaster.

An old proverb says, "A stitch in time saves nine." Today we might say, "Corrective action, taken in time, saves your job and mine."

The only people who benefit from catastrophes are the liability lawyers.

13

Inspecting for Defects

In **Faust**, Goethe wrote, "Man errs as long as he strives." Modern men and women must be doing a lot of striving in the workplace; there is an abundance of errors.

Who should catch those errors, mistakes and defects?

- Production operators?
- Quality control auditors?
- Quality control inspectors?
- Special laboratories?
- Automatic testing equipment?
- Universal testing?

PRODUCTION OPERATORS

The modern trend is toward self-inspection where production operators check their own work. The main advantage is that production operators are close to the problems; when they detect an error they can take corrective action immediately. As an example, operators can improve setups, change targets, sharpen tools, or call for help.

In many instances, operators know what caused the problem. They can remember that defects appeared after a tool started to chatter, a new lot of raw material came in, or a process changed. The time between the event that caused the problem and identification of the defects may be short.

Self-inspection works best when production management considers quality as important as quantity. Most production managers profess this philosophy. Few adhere to it. Their true priorities are shown when they are behind schedule or when they promote subordinates.

Managers who always favor high-volume producers, regardless of quality, should not be in charge of the inspection function.

The main disadvantage of self-inspection is that you don't want the fox guarding the hen house unless someone is watching the fox.

QUALITY AUDITORS

Even under ideal conditions, someone is needed to verify that the system is working. Independent quality control auditors often perform this function. Auditors verify that the operators are following the right procedures when making and checking the product. Auditors may inspect parts, assemblies, and materials as part of the audit, but their main responsibilities involve the system.

The number of auditors needed in a plant with good self-inspection is less than the number of inspectors needed in a plant with no self-inspection.

QUALITY CONTROL INSPECTORS

Quality control inspectors check production against the specifications. In most cases, they take periodic samples and shut down operations when production gets out of control.

The use of quality control inspectors, however, is not necessarily a reflection on production management. Quality control inspectors are also needed when production operators don't have the skills and authority to perform self-inspection.

SPECIAL LABORATORIES

Testing laboratories are used when special skills, equipment, or chemicals are required. As an example, most petroleum refineries have chemistry and inspection laboratories. Chemical analyses are normally conducted by graduate chemists. Physical analyses are usually conducted by technicians. In both cases, they use apparatus that is not appropriate for the production environment. Some equipment is too large, some is too fragile, and some is unsafe outside of specialized laboratories.

Aerospace electronics firms use special laboratories for cycling tests; these involve shock, vibration, and severe environmental conditions. Temperature cycling, as an example, is used to evaluate the product's reaction to a wide range of temperatures. A single cycle may involve cooling the product to $-40°$ F, holding it there for a specified time, and warming it up to the ambient temperature. The cycle may be repeated many times during the test. Then the product is given performance tests to ensure that the cycling didn't damage it.

AUTOMATIC TESTING EQUIPMENT

Automatic testing equipment is being developed to take over many functions currently handled by operators and inspectors. Although automatic testing equipment is generally expensive, it is efficient in high-volume situations; it increases the speed and decreases the cost of inspection. This enables some companies to perform 100 percent testing for critical variables.

Many beer companies use automatic test equipment to ensure that their cans have the right amount of brew. Too much beer wastes resources. Too little beer cheats the customer.

Automatic test equipment identifies defectives. In many instances, it also gets rid of them.

The more sophisticated automatic testing systems also collect, analyze, and display statistical data on the process. This saves many hours of data collection, analysis, and charting. Before buying the data collection option, however, make sure it will give you up-to-date information. Systems have been known to lag production by many hours. Outdated information has little value.

Automatic testing equipment should not do away with conventional testing. Test equipment may drift, get stuck, or go out of calibration.

Periodic conventional testing is needed to identify these problems before large quantities of discrepant products are manufactured and sold.

UNIVERSAL TESTING

One company in Boulder, Colorado, uses production operator testing, quality control testing, special laboratories, automatic testing, and universal testing.

Universal testing brings all employees into the quality control operation. If an operator, inspector, manager, or clerk finds discrepant material being processed, that individual can shut the operation down. This responsibility, however, can't be taken lightly.

REDUNDANCY

At times, management underestimates the complexity of the inspection process; it is not as simple as it seems. As an example, the drawings for a 1-foot rod may call for a .250-inch diameter. Where should the measurement be taken and how many readings should be made?

No rod is perfectly uniform. Measurements taken at one location along its length will probably be different from those taken at another. In addition, a variety of readings will be obtained if the rod is turned 45° each time it is measured. No rod is perfectly round.

In order to get around this variability, many companies require their inspectors to take a number of measurements. If any reading is above or below the acceptance limits, the part is rejected.

Getting the "right" readings is not easy. Variables influencing measurements include:

- The measuring device
- The item being measured
- The person doing the measurement
- The environmental conditions

MEASURING DEVICES

The perfect measuring device has not been invented; every gage has its own limits of accuracy and precision. As a consequence, you must be sure your measuring equipment will meet your needs.

As an example, carpenters, making measurements that have to be accurate to one fourth of an inch, use metal rulers. Machinists, making measurements that have to be accurate to 0.005 inch, use micrometers.

DIFFERENCES IN ITEMS BEING MEASURED

The item being measured is a significant variable. As an example, it is difficult to accurately measure materials that are easily compressed, like rubber. Even with harder materials, the measurement is dependent on the exact spot where the gaging is placed. This is complicated by the fact that you can never position the gage twice in exactly the same spot. Because of this, it is difficult to get repeat readings when making measurements with sensitive instrumentation.

PEOPLE ALSO VARY

With more complex instrumentation, the person setting up the equipment or making the measurement is another variable. No two people have the same dexterity, eyesight, conscientiousness, experience, understanding, or motivation. Laboratory tests, where everything is controlled, often show marked differences in measurements made by two competent people.

ENVIRONMENT

Environmental conditions at the time of the test are also significant. Almost any product will expand or contract when temperatures change. Precise measurements, including those used to calibrate instruments, usually require that environmental conditions be held within prescribed limits.

Environmental specifications might include temperature, humidity, atmospheric pressure, and air contaminants. Temperature changes cause expansion and contraction of parts and gages. High humidity can cause corrosion. You don't want to damage your product while it is being examined. Atmospheric pressure influences sensitive pressure gages. Finally, air contaminants can damage moving parts with tight tolerances like disk drives and gyroscopes; they can also cause corrosion.

GETTING A REPRESENTATIVE SAMPLE

In order to control a process, you must measure the product accurately. You must also take samples that represent the process.

In petroleum refineries, the job of sampling is frequently given to junior operators. Once taken, the sample is sent to the laboratory and analyzed to determine whether the operating unit is running smoothly.

Green production personnel often grab samples by sticking their container under the outlet and opening the valve. Material that has been settling in the line since the last sample goes into the container first.

The proper method of sampling is to drain material out of the line until it has been thoroughly flushed, then to fill the container.

When operating personnel know the unit has been running smoothly and the lab knows the sample was out of specification, problems are sure to develop. Both can be right. In most cases, the problem is with the sample.

If your sample isn't any good, it doesn't matter how accurate your measurements are.

Samples taken from petroleum refinery storage tanks present another problem: The tanks may be stratified. A stratified tank is one whose contents are not uniform.

In order to check on this possibility, refineries usually take three samples from each tank, one each from the top, center, and bottom. A special sampling tool, called a thief, is used to perform this task. The thief is closed, lowered to the desired level in the tank, and then opened so that the liquid can flow in.

Analyses of samples from the top, center, and bottom are used to determine whether the tank is stratified or uniform. If the samples vary too much, the tank is stratified. Material in the tank must be mixed until the samples are more consistent.

When making parts, assemblies, and solid products, the equivalent of a storage tank is called a "lot." In other words, a lot is a unit of production. In some cases it is one shift or one day's production. In other cases a lot is the product a company made between the time it started its current run and the time it started making something else. This is particularly true with job shops, where short production runs are common.

"Lots" of solid products are stratified when the operation is not running properly, just like storage tanks.

IS THE PROCESS IN CONTROL?

Two other common terms used in quality control are "in control" and "out of control." If the operation is running properly, it is called "in control." If it is running improperly, it is called "out of control."

When an operation is in control, the pallets and containers are relatively homogeneous; variations in the product are random. When it is out of control, they are stratified; pockets of discrepant product are scattered throughout the lots.

Sampling enables you to estimate the average quality of a lot. The more items you measure, the better their average approximates the average measurements of the lot—assuming the process was in control at the time the lot was being formed.

Many sampling systems assume the process is in control. This, however, is not always a good assumption. Processes are like hyperactive two-year-old children. They don't stay in control unless special efforts are made to keep them there.

Out-of-control lots will have pockets of defective material resulting from assignable causes. Some sections of the lot will be significantly different from others. The only way an out-of-control lot can be appraised with a high level of confidence is to measure every item in it.

EVALUATING THE PROCESS

Inspection for defects is not a simple process. It involves:

- Taking a good sample
- Measuring it properly
- Understanding the limitations of the sample and the measurements

Without a thorough knowledge of all three, you are apt to draw the wrong conclusions and take the wrong actions.

When your system is in control, don't make changes. Many in-control operations have been thrown out of control when operators made unnecessary changes. When the system is out of control, don't try to wish your problems away. When in doubt get the right answers before you act.

14

Reporting Defects

As Arthur Hays Sulzberger (1891–1968) said, "A man's judgment cannot be better than the information on which he has based it."

It is the quality manager's job to report the truth about quality. It should be unbiased, clear, and concise. It should be expressed in terms that even top management can understand. In other words, avoid technical jargon and local abbreviations when possible.

Members of upper management are not stupid. They won their promotions because they had an edge on their competition. On the other hand, few top managers understand the technical aspects of quality.

Specialists shouldn't try to impress people with their knowledge; they should help others benefit from it. If management learns a little about the technical side of quality assurance, so much the better.

Show management the costs of specific quality problems. This will get their attention.

One company president used to shuffle through papers and handle routine tasks while he was being briefed. Once the briefer started talking about the bottom line, however, the president became attentive; you could almost see the dollar signs blinking behind his glasses.

Executives depend on plant tours, briefings, and reports to find what's happening. They are related.

PLANT TOURS

Managers who walk through the plant every day usually know what's going on; if there are quality problems, they know about them before the quality status reports hit their desks. When problems are seen and discussed on the shop floor, there is little chance for misunderstanding; first-hand observations can't be distorted or screened by anyone.

Disadvantages of daily shop tours include:

- They consume valuable time.
- Lower-level managers feel bypassed. In addition, they are embarrassed when they learn about problems from people they report to.
- The boss may be misinformed. Some people in the shop may be interested in making an impression without getting their facts straight. Others get flustered or intimidated. Many of these problems can be overcome, however, if the manager listens with an open mind and verifies the information with trusted subordinates.

BRIEFINGS

Briefings come in three categories: informal, semiformal, and formal. When managers walk through the work area, they frequently get informal briefings from workers and supervisors. When they get back to their desks, they often request informal briefings from their direct subordinates, hoping to verify what they heard.

In time, their subordinates will make tours of their own to reduce the risk of learning about problems from the boss. As a consequence, both the executives and the subordinates become more aware of what is going on.

Semiformal briefings usually involve requests for information on a specific project with little or no time to prepare. Astute briefers, however, usually have the data for a formal briefing at their fingertips. Their briefings are essentially formal, although their charts may not be up to date.

Formal briefings are much more demanding. They should:

- Be brief
- Be clear
- Cover all important points
- Be supplemented by transparencies or slides

For effective briefings it is best to have more backup then presentation. Critical material must be presented. Additional material should be kept in reserve, to be used when answering questions.

The level of brevity depends on the audience. Some executives start fidgeting after five minutes. Others last for over an hour without losing interest.

Clarity involves talking in plain English whenever possible. Good briefers cover technical material without using words and abbreviations the audience doesn't understand.

Good briefers know how to emphasize important points without omitting critical parts. They also know how to condense the last 30 minutes of a presentation into a 5-minute summary as soon as the boss starts to fidget.

When giving a presentation, study the eyes of your audience. This adds a personal touch. It also gives you a warning when your boss' eyes start to glaze. If this happens, don't panic. Summarize. The boss may have had a rough night.

Slides and transparencies are an important part of most formal briefings. Many people use them. Few people use them well. The best slides and transparencies are simple, readable, and self-explanatory.

Four simple illustrations are better than one that is too complex; only the originator may know what it says.

Confusing graphs include those that have three or more different curves wandering off in different directions. Most of these illustrations try to compare different years or products on a single chart.

Experienced illustrators reduce confusion by using overlays. Then, each curve can be shown and discussed by itself. If comparisons are needed at the end of the presentation, the individual charts can be overlaid or placed on top of each other. Then all of the curves will be projected on the screen at the same time. This technique postpones the confusion caused by complex charts until after the audience is familiar with the individual data. This way, the attendees have a fighting chance at understanding the presentation.

Readability is critical. Some inexperienced briefers tend to cram too much information into a chart. In the end they are the only ones who can read it.

A good rule of thumb is to have between five and ten lines on the chart. Fewer lines make the charts inefficient. More lines make the print so small that it can't be read from the back of the room.

With transparencies, the printing should be at least 3/8 inch high. Half-inch lettering is better. With flip charts, the lettering should be at least 1 in. high. Why is there a difference? Transparencies get magnified; flip charts don't.

Computer printouts and typewritten data don't meet either criteria. If they are needed, include them in handouts. People don't want to strain their eyes squinting at illegible data. Those who try fall behind and can't follow the speaker.

WRITTEN REPORTS

The only firm rule about written reports is "be flexible." Most written reports have to conform to the format set by the boss.

Members of the process department of a West Coast oil company used to groan every time they had a new supervisor. They knew the report format would change as soon as their new boss learned how to spell the department's name.

Rules that are generally accepted, however, include:

- If the information can be handled via the phone, don't bother writing the report.
- Follow the format recommended by your boss.
- Start with a summary. Include enough information so the recipients will know whether to scan, read, or study the rest of the report.
- Avoid technical jargon and abbreviations. Diehards who try to force others to learn their pet phrases, technical terms, and abbreviations lose their ability to sell ideas. It is too easy for the upper echelon to dismiss them as eggheads and ignore their conclusions, no matter how worthy.

WHAT INFORMATION?

What information should the boss be given? Seven areas of quality that should be contained in recurring reports and briefings include trends and developments in:

1. The current rate of field returns and complaints
2. Internally generated defects
3. Improvements in inspection and testing
4. Production deficiencies affecting quality
5. Design deficiencies affecting quality
6. Raw material deficiencies affecting quality
7. Quality costs

THE CURRENT RATE OF FIELD RETURNS AND COMPLAINTS

This information is critical; it gives feedback on whether the company is doing a quality job. If possible, identify how many items were defective when shipped, damaged by the customer, or stored too long.

Clerical and miscellaneous errors involving customers should be reported also. Improper billings and incorrect shipments are particularly meaningful.

Backup information should include lot numbers and dates of shipment. These data are part of all good quality control systems; they help identify what caused the defects.

In some cases, there will be a large number of returns which have nothing wrong with them. These data should be presented. The returns may reflect poor communications with the customer, a problem that should be corrected.

Until field returns approach the vanishing point, management should be told where the defects arose and what is being done to prevent recurrence.

INTERNALLY GENERATED DEFECTS

These defects should be identified by type of defect, probable source, disposition, and corrective action. Summaries can be effectively reported to management by use of bar graphs called Pareto charts.

A simplified Pareto chart is shown in Figure 14.1.

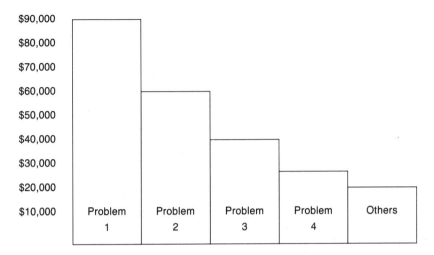

Figure 14.1 Monthly cost of defects by defect. Problem 1 = printing, problem 2 = coating, problem 3 = metal forming, and problem 4 = contamination.

In the simplified case shown in Figure 14.1, printing errors cost the company more than any other defect. The second most expensive problem is coating; the third is metal forming; the fourth is contamination. This information tells management where the most expensive problems are located; it enables management to concentrate its problem-solving talents in areas containing the largest potential savings.

Many Pareto charts emphasize the number of defects instead of their costs. Management, however, normally stresses dollars, not defects.

INSPECTION AND TESTING IMPROVEMENTS

It is easy to stick with an inspection and testing program that has worked for ten years. Nothing, however, is perfect. Review the programs.

Study field returns and internally generated defects. They may help identify where improvements should be made.

If printing is the greatest source of defects, for example, the company should look for improvements in that area. Can problems be identified quicker? Can testing techniques be developed to check incoming inks before they go to production? Can a first-in/first-out inventory system be implemented to reduce the risk of shelf-life (storage) problems?

As long as existing programs haven't reduced critical and major defects to the vanishing point, there is still room for improvement.

PRODUCTION DEFICIENCIES

Pareto charts give management an idea of which production defects are causing most of the problems. Every third or fourth presentation by the quality manager should cover these problems in detail. Information presented should include:

- How much these defects cost the company
- What is being done about them
- What additional help is needed from top management

The quality manager should cooperate with the production manager and give production credit for corrective action that has been taken.

DESIGN DEFICIENCIES

The quality department should be responsible for relating specification tolerances to process capabilities. It should also help evaluate the causes for warranty repairs, return products, and customer complaints. Experience in these areas gives the quality department the background to help identify and resolve design problems, even though some designers may object.

The quality manager should present information on design deficiencies and explain what is being done to correct them. Before the presentation, however, the quality manager should discuss the report or briefing with the design manager. This will reduce the risk of confrontations involving top management. It will also promote cooperation.

By working with design, the quality manager ensures that credit is given for corrective action that has taken place. If the problem cannot be resolved without funding or other assistance from top management, that information should also be presented.

RAW MATERIAL DEFICIENCIES

Being responsible for receiving inspection, the quality manager is often the first to know of raw material problems.

The quality manager should get input from the purchasing and design organizations concerning action taken to correct the raw material problems. Corrective action may involve preshipment testing of incoming raw materials. This is especially desirable when shipping schedules are tight and unacceptable materials have been received in the past.

Preshipment testing involves making small test runs with each incoming lot of raw material. If the material proves unacceptable, it can be returned to the vendor and replaced before shortages develop. If the material runs smoothly, production can schedule with confidence that defective material won't force it to look for a replacement lot.

QUALITY COSTS

Quality costs are an important part of most communications with top management.

Almost everything in the company involves quality and quality cost. The quality manager must work closely with the accounting department to develop accurate figures. The main categories include:

- Quality appraisal
- Defect prevention

- Scrap and rework
- Diagnosis or trouble-shooting

Quality appraisal includes the inspection and testing functions, whether they are performed by the production (self-inspection) or the quality organization. Appraisal also includes the equipment and supplies needed to determine whether parts and materials are acceptable.

Defect prevention is relatively new to the quality scene. Few companies give it the attention, funds, and personnel it needs. They remember the old saying, "An ounce of prevention is worth a pound of cure," but they don't apply it to themselves. This includes those who work with marketing, sales, design, manufacturing, packaging, and shipping. They are all in a position to prevent defects. Some do; some don't.

Scrap and rework are both expensive. They not only waste production time and materials, they also waste the time of the engineers and managers who decide how to dispose of discrepant material. These expenses are quality costs.

Diagnosis and trouble-shooting involve finding the causes and remedies for quality problems. In some companies, the trouble-shooters do little more than put out fires. The modern trend, however, is to anticipate and prevent problems. As a consequence, these costs are closely aligned with prevention programs.

Most quality costs are presented to management as dollar amounts. When cost trends are being presented, however, it is best to show them as percent of production costs or percent of sales. This approach helps reduce the fluctuations that come with changes in production volume.

Reports and briefings should not cover all seven areas at one time. Most briefings, however, should include field reports, customer complaints, internal defects, and quality costs. The other areas should be covered as problems arise and time permits.

Inspection and testing improvements can be addressed at one briefing, production deficiencies at the next, and design deficiencies at the third. Then, the cycle can be repeated.

The quality manager shouldn't wait for management to request this information. It should be contained in routine reports and briefings with a strong emphasis on costs. Top-level managers understand dollars much more than they understand statistics and quality control. Anyone who wants to get through to them must speak their language.

15

Key 4: Coordination

According to the Doubleday Dictionary, coordination is "the harmonious action of the various parts of a system." When paraphrased and applied to industry, that definition might be, "Let's quit cutting throats and get the job done."

Virtually everyone in the company has an effect on quality, some directly, some indirectly. No organization can be confident of its quality unless everyone is contributing.

Coordination is needed. If quality is everyone's responsibility in general and no one's responsibility in particular, crises develop. Critical requirements are overlooked when each department thinks someone else is "minding the store."

In some companies, coordination takes place in the president's staff meetings. The boss makes a decision and the staff respond with a coordinated "Yes, Sir" or "Yes, Ma'am."

In formal organizations, coordination involves having all matters affecting other departments go to the department head. Then they are transmitted from department head to department head—or from department head's secretary to department head's secretary. Eventually, they are passed down to the appropriate individual.

Less formal organizations have a free and easy interchange of information where everyone works with others regardless of departmental affiliation. Upward, downward, and lateral communication is encouraged as long as the privilege isn't abused. Company presidents don't normally want to discuss petty details with entry-level personnel unless the problem is important to both of them.

The key to successful control of product quality does not lie in whether management is formal or informal. It depends on whether coordination is adequate. Top-level managers must choose the kind and degree of coordination they want. Then they must ensure that coordination takes place.

The level of coordination that is needed is influenced by how the product is used and how complex it is. The most hazardous situation is when the product is new, complex, and involves state-of-the-art technologies.

Close coordination is also needed when introducing a new model of a complex product. In these cases, it is easy to assume that no special effort is needed to control quality. This assumption is wrong.

In the aerospace industry, for example, some of the biggest problems have come from "nonfunctional" changes in known products. People proposing the modifications claim, "These changes couldn't hurt anything." But they often do. Oddly enough, many nonfunctional changes have been instituted to correct problems that were less damaging than the ones they created.

Less coordination is normally needed when products are simple and familiar to everyone. Plastic glasses are a good example. They look innocent. They can, however, create problems with sharp edges or unstable chemicals. In addition, internal losses due to scrap and rework can eat up the profits.

The most effective coordination comes when the work of different departments is effectively tied together. Then the whole becomes greater than the sum of the separate parts. Reductions in scrap, rework, customer complaints, and product recalls become greater than anything achieved by the best efforts of uncoordinated departments working independently.

Most complex products, using new technologies, can be improved by quality teams. With these items effective coordination is critical. Properly organized quality teams can:

- Anticipate future problems
- Identify existing problems
- Coordinate their resolution

Members of the quality team should ensure that their departments are dedicated to preventing quality problems. Each team member should also become a devil's advocate, ferreting out quality hazards and weaknesses.

Under ideal conditions the quality team will anticipate new product problems during the design phase. The earlier they catch a problem, the less it costs. As many stock market advisors have said, "Take small losses early but let your winnings run."

Quality teams are made up of specialists with a variety of skills. They can identify problems that no one individual would anticipate. The coordinated action of a quality team reduces the probability of each department blaming another when things go wrong.

Quality teams should include the heads of the following functions:

- Quality
- Marketing
- Finance
- Design
- Manufacturing
- Purchasing
- Field services
- Packaging and shipping

The quality department provides skills for measuring quality, analyzing quality data, identifying process capability, setting specifications, writing procedures, designing experiments, and providing statistical studies. Quality is their name.

Marketing provides skills for identifying customer desires, proposing new products, conducting marketing surveys, estimating markets, and pricing products. It can also profile probable customers and identify characteristics that are important to the buyers.

Marketing has the closest contact with consumers. If the time is right for a new product, marketing should be the first to know. If problems develop with an old line, it should be the first to get feedback from the customer.

Finance and accounting enable the team to keep track of where the money can come from and where it goes. Before committing to a new product, they need to know the estimated costs of design, manufacturing equipment, raw materials, staff, training, facilities, testing equipment, and packaging. They also need information on probable costs of scrap, rework, warranties, and field service.

Design is responsible for converting the dreams of the marketing department into drawings and specifications that can be used by manufacturing. These drawings will specify materials, dimensions, tolerances, and performance requirements. The drawings should also identify critical characteristics that might result in hazardous or unsafe products. In addition, they should identify major defects that can cause product failures.

The head of manufacturing can provide the quality team with information on what can and can't be produced with existing equipment and machinery. As an example, it is easy for a design engineer to specify ± 0.0001-inch tolerances; it may be difficult or impossible for manufacturing personnel to meet them.

Manufacturing will also have information on materials and suppliers that caused production problems in the past. Some of this information may not be available to the quality organization; the problems may have appeared after the shipments cleared receiving inspection.

The head of purchasing will have information on suppliers who can make the necessary parts, equipment, and raw materials. In addition, purchasing should be the first to know if existing vendors are having problems. As an example, they may find that the only chemical plant producing one of the proposed raw materials has burned down. Then a more available material must be substituted.

The head of field services will have information on problems happening in the field. Although this information generally relates to current products, it will also apply to new or modified items using similar raw materials, subassemblies, or spare parts.

The head of packaging and shipping will have knowledge about the techniques and materials available for protecting the product after it has been manufactured. This includes the best protective materials, the appropriate shipping routes and the available common carriers.

Although the quality team should have representation from most major divisions of the company, the president shouldn't be included.

The quality team can operate most effectively when ideas can be tossed around freely without members having to worry about what the boss thinks. The team needs the unqualified support of the president, but not the presence.

Membership on the quality team should be limited to heads of major company functions. At times, however, substitutes will be inevitable. When substitutes are sent to meetings of the quality team, they should:

- Be able to commit their organization
- Have quick access to the official team member
- Have the agenda several days before the meeting

All quality team members and their delegates must have the authority to commit their organizations. The team becomes ineffective if substitutes agree to collective action only to have their position reversed after their boss gets briefed. As a consequence, "Theory X" managers, incapable of delegating authority, shouldn't send substitutes.

Substitutes must have quick access to the official team member. There are times when even the best substitutes have to consult with bosses before making a difficult decision. These consultations, however, shouldn't be preceded by a three-hour wait in the outer office while grievance committees, EEOC representatives, and others parade in and out.

Quality team meetings should have agendas and all attendees should have access to them several days before the meeting. This includes substitutes. Meetings without agendas tend to wander. Attendees who are unprepared are seldom willing to make major commitments.

How often should the quality team meet? It is top management's responsibility to see that the quality team meets often enough to ensure the following:

1. When production starts, scrap and rework levels are acceptable.
2. Production schedules are achieved as planned.
3. Delivery dates are met.

4. Customer returns and complaints are close to the point of diminishing returns.
5. The likelihood of liability claims is essentially nonexistent.

For new products, there should be several meetings for each stage of the planning. This starts with the marketing phase and continues until the product is in the marketplace.

During the preproduction phases, problems can be resolved before they become serious. At times, however, pressures build; the company wants the product in the stores ahead of the competition. This is a good objective but it is no excuse for selling shoddy products. Premature entry into the marketplace can cause companies to lose customers or even their businesses.

In order to ensure top management's involvement, the quality team should prepare a one-page summary of projects agreed on at each meeting. The sheet should also include who is responsible for each project and when it is due. Copies should go to all members of the team to remind them of their commitments. Information copies should go to top management to keep them informed. It is difficult to get top management's support if it doesn't know what's going on.

SUMMARY

Quality teams improve coordination within the company. They also require a lot of meetings and tie up key people. What benefits can be gained by this overhead? Would you like to reduce scrap and rework by 90 percent? Would you like to reduce customer complaints and returns by an equal amount? Would you like to reduce or eliminate recalls? Would you like to protect your company from liability suits? Then improve coordination. Coordination can help you achieve these goals. Quality teams can help you coordinate.

16

Making Coordination Work

Coordination doesn't come easily. You have to work at it. So does top management.

Top management often retards coordination by giving departments too much independence or failing to give them direction.

INDEPENDENCE

Departments that are too independent resemble adjacent houses where the occupants are not speaking to each other.

For example, suppose, the marketing and salespeople are in the first house. When members of the marketing clan identify a new product that is in demand, they write the information on a scrap of paper. They make the note into a paper glider and toss it out their window towards the design clan's house next door. With luck, the designers will pick it up.

From then on, no communication takes place between the two departments. If instructions in the note are not clear, the designers draw their own conclusions and proceed. It is too much bother to contact people in the marketing house. "After all," they say, "marketing doesn't know anything about design, and their phone may be out of order anyway. Besides, it's raining outside. Who wants to get wet?"

When the designer clan completes the drawings and specifications, they open their window and communicate with the manufacturing engineering house on the other side using another paper glider.

The manufacturing engineering clan then analyzes the tooling and facility requirements without talking to either the designers or the marketing people.

The project passes through the finance, manufacturing, inspection, testing, packaging, and shipping clans and eventually reaches the customers. Much to everyone's surprise, the consumers balk. This wasn't what they wanted at all. Why?

The marketing department did a thorough market survey. The design department prepared drawings and specifications that matched its concept of what marketing wanted. The manufacturing engineering department recommended the proper tooling and facilities based on the note they received. The finance department made sure the pricing was competitive. The manufacturing department complied with the drawings and specifications. The quality department made sure that no defectives slipped through.

Members of each department did their jobs as they saw them. The only problem was they weren't talking to each other. The end product didn't come close to meeting the desires of either marketing or the customer.

Now everyone started talking to each other, pointing fingers, and trying to place the blame on someone else.

The right time to communicate and coordinate is at the very beginning, not after the product is in the market place.

Companies that do not communicate and coordinate during the preproduction period are playing a parlor game they learned as children. Do you remember forming a large circle and having one person whisper details of a dramatic event to another? The story was passed from one person to the next, all around the circle. By the time it returned to the source, it was never the same as it was when it started out.

If a company doesn't have good communication and coordination, top management will not get the products it wants. Neither will the customer.

Communication, cooperation, and coordination have no start when a new product is being conceived and continue until the company stops making it. In addition, people have to document what they do.

A story is told about a production supervisor who asked a design engineer for help and the engineer cooperated. They both went into the shop and discussed the problem. Once the designer saw that the manufacturing equipment wasn't capable of making parts that met the design requirements, the procedures were modified, design tolerances were relaxed, and design details were changed. In other words, compromises were negotiated by two individuals. In the heat of the bargaining, the designer forgot to document the changes. The part was modified but the drawings were not. Eventually, the final assembly supervisor found that some of the components didn't fit together and some of the finished units didn't work. A screening of the inventory showed that some of the parts were made the old way and some were made the new.

When this happened, the design engineers were told to stay out of the plant and the manufacturing supervisors were told to stop talking to the design engineers. The subsequent disruption of communications hurt the company.

To prevent recurrence of the problem, everyone was told that all modifications had to be requested via formal change notices and all change notices required executive approval. The pendulum swung the other way.

Did this really happen? Probably, and often. For best results, coordination and cooperation must involve management as well as subordinates.

MANAGEMENT DIRECTION

Management-directed coordination provides the following:

1. Improved planning
2. More defect prevention
3. Better use of past experience
4. Happier customers

IMPROVED PLANNING

Planning helps a company determine whether new equipment can produce new parts to specification before the machines are bought—not after production has started and the company is committed to unacceptable machinery.

One company installed a multistage machine and put it into production without testing it. High defects were noted immediately, and the equipment couldn't be adjusted to correct the problem. The machine had to be completely rebuilt before it produced acceptable parts. Failure to plan wasted time as well as money.

DEFECT PREVENTION

Defect prevention is the main function of modern quality control, but many companies are not up to date. When mass production began, the main way of keeping bad products from the customer was to inspect the items and weed out the defectives. Today, a century and three quarters later, many companies still emphasize the inspection and removal of bad parts. They don't concentrate on making them right the first time.

Experience can help. As Oliver Wendell Holmes, Jr. (1841-1935) said, ". . . a page of history is worth a volume of logic." What is history if not experience?

PAST EXPERIENCE

When problems arise, most companies have volumes of experience to help them find the solution. Unfortunately, they often fail to take full advantage of their historical files and the experience of their staff.

One quality engineer, aware of his limited exposure to plant operations, found an experienced supervisor to work with him. The engineer supplied the technical background; the supervisor supplied the experience. Between them, they solved so many of the plant's problems that both were promoted in less than a year.

PLEASED CUSTOMERS

Pleased customers lead to profits, and profits lead to successful companies. Coordination, supervised and directed by top management, can result in wise selection of raw materials, efficient production, low scrap, minimal rework, happy customers, and pleased stockholders.

ALTERNATIVES

What are the alternatives to coordination?

Imagine a football team in which no player knew what to expect from others on the squad. Imagine fielding eleven prima donna grid stars who had never practiced as a team. Would they know what kind of play was indicated by the signal caller? Could they beat a solid team that had played together all season?

In team sports, almost everyone is aware of the importance of teamwork, coordination, and communication. In business, the awareness is more subdued. Departments may not know what to expect from each other. They are seldom trained to work as a team. It is no wonder that the newspapers carry so many stories of product recalls and liability suits.

QUALITY TEAMS

Quality Teams can do a lot to promote teamwork. As an example, they can ensure that preliminary drawings are reviewed by all functional organizations of the company.

Marketing should be satisfied that potential customers will be pleased with the product design.

Design should work with marketing, quality, production, and field services to ensure that the product is producible and that marketing is getting what they asked for.

Finance should work with marketing, design, and manufacturing to ensure that pricing makes the product competitive but profitable.

Production engineering should work with design and manufacturing to ensure that the product is designed to take advantage of efficient machinery and processes.

Purchasing should work with production engineering, design, quality assurance, and manufacturing to ensure that raw materials, parts, and machinery specified by the engineers are readily obtainable at reasonable prices.

Personnel should work with manufacturing, design, and quality control to ensure that training will be available when new employees, operations, or machinery are required.

Quality Assurance should work with design and manufacturing to ensure that inspection planning will detect critical defects without slowing down production excessively.

Packaging and shipping should work with the product development, quality, and laboratory departments to see that tests are run to verify that the product will reach the stores without being damaged.

Field services should provide purchasing, quality, and design with historical data on problems they found with similar products and materials.

Changes, submitted by the reviewers, should be studied by design. Changes that seem reasonable should be implemented. Those in doubt should be negotiated with organizations who submitted them.

Once coordination is complete, the drawings and specifications should be changed to comply with the consensus. If consensus is not possible, the final decision will have to come from top management.

Revised drawings and specifications should be resubmitted for a second preproduction review. Then the process should be repeated.

BENEFITS

If this procedure is followed:

- Tolerances will be more realistic.
- Everyone will get credit for successes.
- Witch hunts will become unnecessary.

After getting the coordination of all functions in the company, designers don't have to add a margin of safety to their tolerances. They can be loose enough to enable manufacturing to meet them and tight enough to ensure a good product. The best specifications define an acceptable product without being so tight that designers accept discrepant parts for further processing.

When everyone gets credit for successes, they consider themselves part of the team. This procedure reduces the "us" vs. "them" philosophy of confrontation. When everyone has a stake in the success of the product, they all try to make it a winner.

When witch hunts are no longer the company's favorite recreation, design engineers don't have to worry about being burned at the stake if anything goes wrong. They can quit specifying impossible tolerances.

Once manufacturing personnel have a say in the design, they are less apt to blame their problems on the drawings and specifications.

Coordination is not easy. Top management must see that preproduction design reviews are held and that they are informed of the results. Members of the quality team must take their reviews seriously.

During design review in a West Coast aerospace company, one of the design engineers requested extremely tight tolerances for one of the specifications. When challenged by the manufacturing engineer, the designer's response was, "Don't worry about it. If the specifications are too tight, we can change them after we get into production." About that time, the government imposed a "no-change" policy, making it almost impossible to relax the tolerances. Fortunately, the manufacturing engineer did not give in to the designer.

When design reviews are conducted properly, specifications are based on what is needed. They shouldn't be too tight or too loose.

Is all this coordination worth while?

One company had .35 percent customer returns before implementing preproduction design reviews. Their customer returns dropped to .009 percent the year after coordination procedures were implemented. Scrap, rework, and customer complaints were also reduced.

Preproduction coordination procedures can yield big, measurable payoffs. They can also help a company stay in business by drastically reducing product recalls and liability suits.

With some states accepting the philosophy of strict liability, companies must do all they can to protect themselves when designing and manufacturing products.

What is strict liability? It is the contention that plaintiffs don't have to prove negligence in the manufacture or design of the product. You might say the manufacturer is guilty until proven innocent.

SUMMARY

Do you want your company to survive and prosper? Then use the three "C's" of survival: coordination, cooperation, and communication. In other words, make coordination work.

17

Coordinators: Who Will Lead?

During a television address, President John F. Kennedy (1917–1963) said, "It is time for a new generation of leadership to cope with new problems and new opportunities. For there is a new world to be won."

What greater world is there than one that's defect-free?

Who should lead this war against defects? The head of the quality team? If so, who should lead the team?

"Quality" is in the title but that does not dictate who the leader should be. Personality, skills, background, and connections should be evaluated when making the choice.

Characteristics needed by the leader of the quality team include sales ability, technical knowledge, and support at the top.

SALES ABILITY

Sales ability requires a good personality. In the industrial sense, it also requires cooperativeness, persistence, an inquiring mind, thoroughness, clarity of expression, calmness under pressure, and dependability.

Sales ability is not limited to selling hardware. It also involves selling ideas, reputations, and people. Sales ability is a necessary part of any job.

116

Babies sell their mothers and fathers into taking care of them by being cute. Lovers sell prospective mates on marriage by promising them the moon. Engineers sell their projects by writing impressive reports.

The head of the quality team, however, has the greatest sales job of all. The person selected must be a leader among leaders and a salesperson among salespersons. Without those qualifications, the leader will have difficulty coordinating the activities of other successful managers.

Is the head of quality a great communicator? Good at selling ideas? Capable of explaining complex analyses in terms that top management can understand? Adept at summarizing six-month projects in one-page reports? If so, then he or she is a good candidate. If not, keep looking.

How many times have you seen competent experts who had the respect of everyone in the company but couldn't sell ideas? Some had the personality of a Genghis Khan with festering boils. People fought their ideas rather than knuckling under, even if they knew the plans were good.

How many times have you seen an idea rejected because the originator tried to dominate the group or couldn't talk in nontechnical terms? The suggestion might have solved the problem but management wouldn't accept solutions from people it didn't like; nor would it accept proposals it didn't understand.

An engineer in a West Coast rocket plant had that problem. He knew his job but he used too much technical jargon. His inability to communicate with production looked unsolvable until a propellant line shut down because burning rates were out of specification.

The line foreman tried everything without success. As a final act of desperation, the foreman called the engineer and asked for help.

The engineer was briefed over the phone. Ten minutes later, he recommended a different lot for one of the chemicals used in the formulation. The foreman found the requested lot and used it in the next batch of propellant. The batch was good. From then on, the two were on speaking terms, although the engineer had to be careful he didn't use too many technical words.

Poor communication and bad sales technique had been a problem. It was resolved by effective trouble-shooting. When people know where they can get results, they are often willing to put up with a few big words—but not too many.

TECHNICAL KNOWLEDGE

The head of the quality team must have enough technical knowledge to intelligently evaluate what others say. As Goethe (1749-1832) said, "Nothing is more terrible than ignorance in action."

A persuasive individual with executive backing and inadequate knowledge can do unlimited damage.

The head of the quality team should know marketing well enough to understand:

- Were the marketing surveys realistic or superficial?
- Did the sampling represent the population or was it biased?
- Were the survey takers given too much freedom in determining who would be interviewed?

Sampling is as important in market surveys as it is in quality control.

How much valid information could you get on the marketability of Lincoln limousines from the residents of skid row? Could you conduct unbiased interviews on the goals and desires of welfare patients by visiting the nicer parts of Beverly Hills?

The head of the quality team should also have some understanding of drawings, tolerances, and specifications. Anyone expecting to achieve ± .001-inch tolerances with a hacksaw would be slightly out of place. Anyone expecting to hold ± .0001-inch tolerances with a 20-year-old lathe that hasn't been maintained would also have problems.

Some knowledge of production is an additional plus. If the company runs a large petroleum refinery, for example, you wouldn't want someone trying to save money by recommending a one-shift operation. It takes more than a shift to start up a large refinery, and more than a shift to shut it down.

Knowledge of statistics is also helpful. Quality teams are usually involved with masses of data. The leader of the team should know whether the data is analyzed properly and whether the raw data are reliable. Even the most sophisticated statistical analyses have little value if the data are no good. As they say in the field of computers, "Garbage in means garbage out."

Other statistical questions include: What statistical tools are available? Which ones are appropriate for a specific job? What are their advantages and disadvantages? What are their limitations? What assumptions are involved?

Can the candidate:

- Put data into a form that everyone understands?
- Supervise storage and retrieval of large masses of data?
- Evaluate which data to throw out when the computer data base gets overloaded?

Information on shifts, machines, suppliers, raw material lots, and preproduction tests may help identify the cause of defects. This data must be stored safely and be retrievable. There are few things more frustrating than having data in the computer and not being able to access it when necessary.

Knowledge of people is also beneficial. What do they know? What don't they know? What do they think they know but don't? What are their strengths and weaknesses? What are their biases?

Biases to avoid include:

- The contention that quality assurance is an unnecessary evil.
- The belief that if you take care of quality, production numbers will take care of themselves.

SUPPORT AT THE TOP

Quality team projects, like most other programs, become exercises in futility when they don't have the support of top management. One way to evaluate top management's commitment to the team is to look at who is selected to lead it. Then, look at how management treats that individual afterwards.

To be effective, quality teams must be headed by individuals with proven records of accomplishmnent and strong support from upper management. They have to get into an executive's office without getting bypassed by the boss' favorite sons and daughters.

How many times have you sat outside an executive's office, only to have some brassy individual walk into the office ahead of you? Those who bypass others create resentment. Bosses who allow bypassing waste their subordinate's time.

The ideal leader of the quality team will not fit either classification. Appointments set up by that mythical individual will be kept by the top echelon, without the team leader being pushy, stepping on toes, or creating enemies. People with many enemies have a difficult time selling programs.

Does anyone fit all of the above criteria? Probably not. It is nice, however, to know the qualifications of an ideal candidate.

SHOULD QUALITY ASSURANCE HEAD THE TEAM?

The name "quality team" implies that it should be run and controlled by the head of quality. There are reasons for and against this approach.

Reasons for selecting the head of quality include:

- Knowledge of statistics, data analysis, and quality assurance are important. The head of the quality function usually has this background. If not, the company probably has a low regard for quality and the quality team. It is apt to end up as window dressing, not a vital force in the company.
- The head of the quality function should already have extensive experience with sales techniques. This is often gained by trying to sell manufacturing on the concept that high production rates and good quality are not necessarily synonymous.

Reasons for not selecting the head of quality include:

- When the quality team is headed by the chief of quality, it is easy to look at the team as another addition to overhead. If the program fails, the head of quality becomes an expendable scapegoat.

- If the program succeeds, the head of quality survives another annual review, thanks to the help of others who think they deserve the credit.
- The head of quality is often accused of slowing down other functions, interfering with their operations, and getting in the way.
- If a leader is selected from outside of quality, the team members can see that quality is everyone's business; it is not a boondoggle that inflates overhead.
- The head of quality may not be strong in sales technique. The program must be sold. It will not sell itself.
- The head of quality may not have the political base needed to get support from top management.
- The head of another function may be more qualified.
- The price of unqualified support from other functions may be the prestige of leadership.
- The head of quality may not be technically qualified.

One of the biggest problems top management has in selecting a quality manager is evaluating the individual's technical competence. When top management knows little about quality assurance, it is difficult for them to determine whether an applicant is qualified in that area. Unfortunately, few executives will admit that they don't know as much about quality as anyone else in the plant.

The price of accepting the leader's role is also high. Top management must make it clear that the leader—and the members of the Team—have a stake in its success.

HOW CAN THE QUALITY MANAGER HELP?

Team leader or not, the head of quality should be a team member and a valuable resource. Services that the quality manager can provide include:

- The quality manager can help design engineers identify which quality characteristics are important and how they should be inspected. Quality might also have input on preproduction testing, simplified assembly procedures, and methods of keeping parts from being assembled backwards.

- The quality manager can help production evaluate the capability of its machines and the performance of its workers in the quality area.
- The quality manager can help purchasing evaluate the ability of suppliers to produce acceptable parts and materials.
- The quality manager can help packaging and shipping evaluate whether the products will reach the customers in good condition.
- The quality manager can help marketing evaluate customer surveys.
- The quality manager can also help field services identify which warrenty problems require corrective action.

SUMMARY

The war against defects should be coordinated by an outstanding salesperson with excellent technical knowledge and firm support at the top.

If the head of quality meets these requirements better than anyone else, that individual should have the assignment. If not, the appropriate manager with the best qualifications should be chosen.

Defect prevention is not a parochial assignment. It's everyone's job.

18

Key 5: Effective Problem Solving

The old-fashioned approach to problem solving followed the maxim, "Don't just sit there; do something." The theory was, "If you make enough changes, you are sure to stumble on the right one—eventually."

Today, problem solving is more systematic and efficient. We don't rev up the engine before deciding where to go.

Effective problem solving can be divided into:

1. Problem prevention
2. Problem identification
3. Problem resolution

Why put problem prevention first? If your problem prevention program is good enough, you don't have to worry about the other two.

PROBLEM PREVENTION

The best time to solve problems is before you go into production. Problems caught in the design or prototype stages are relatively inexpensive. Production is not stopped, scrap is limited, and only a few parts are involved.

Problem prevention techniques covered in previous chapters include:

- Training top management
- Training everyone else

- Designing quality into the product
- Testing prototypes
- Working with customers
- Working with suppliers
- Forming quality teams

Statistics provides additional tools for effective problem prevention. The statistical approach, however, is normally used in conjunction with methods already covered.

Statistical tools useful for problem prevention include:

- Process capability studies (Chapter 20)
- Ends tests (Chapter 21)
- Other tests for comparing differences (Chapter 22)

These analyses are preceded by a chapter on basic statistics (Chapter 19) so you won't have to go into the subject "cold turkey."

Details on complex statistical analyses are omitted in order to preserve the sanity of the average reader. References to books containing more information in these areas are given at the end of this chapter.

Process capability studies help design personnel to match their specifications with production's ability to meet them. Therefore, they can be classified as tools for preventing problems. If design doesn't set realistic tolerances, production will not be able to meet them. Then scrap and rework are inevitable.

When process capability studies show that the specifications are too tight, design must change the requirements or justify improvements. In most cases, those improvements involve buying better equipment. In some instances, however, additional training or improvements in the old equipment will do.

Ends tests are relatively simple; they compare processes or materials to see which is best. New is not always better. Imagine changing to a cheaper material, only to find that scrap and rework double.

The Tukey-Duckworth test and the paired T-test provide alternative techniques for comparing materials and processes. The

Tukey-Duckworth test requires few calculations and can be used by technicians and managers as long as the data is checked by a statistician.

The paired **T**-test is a little more complex, but computer programs are available to do the calculations for you.

PROBLEM IDENTIFICATION

Problem prevention procedures are never perfect. As a consequence, they will have to be supplemented by exercises in problem identification and resolution.

In some companies, repetitive problems contribute to scrap and rework without anyone being aware of them. It is hard to correct problems when you don't know they exist.

Statistical tools used to help identify whether there is a problem include:

- Frequency distribution charts (Chapter 23)
- Control charts (Chapter 24)
- Precontrol (Chapter 25)

Frequency distribution charts are often called histograms. They are effective tools for determining whether major changes are occurring in a process.

In addition to helping identify problems, frequency distribution charts often show whether new process capability studies are needed. They should be used to supplement process capability studies, however, not replace them.

Frequency distribution charts normally use readily available plant data; they help you identify how the process is actually run. Process capability studies are run with tight controls; they help you identify how the process could be run under ideal conditions.

Control charts help identify whether the process is running smoothly. Control charts are based on day-to-day inspection data developed in the production area. They are effective in spotting trends, unusual events, and potential problems.

Precontrol is a process control tool developed by Rath and Strong. It helps operators control their own processes. It also provides early warnings when the processes start going out of control.

Once a problem has been identified, you can move to the problem resolution phase.

PROBLEM RESOLUTION

Approaches to problem resolution include:

- Pareto charts
- Cause-and-effect diagrams
- Brainstorming sessions
- Flow charts
- Divide-and-conquer techniques (Chapter 26)
- Regression analyses (Chapter 27)
- Designed experiments (Chapters 28–29)

Once you have identified the existence of problems, you may conclude you are in a swamp fighting alligators. The question then arises, which one should you take care of first? If your existence isn't being threatened by a job-related alligator, take time to plot a Pareto chart like the one in Figure 14.1. It normally pays to take care of the few critical problems that are costing the most. As those are resolved, you can attack the many that are trivial, if time and funds permit.

It seldom pays to mount million-dollar campaigns to resolve nickel-and-dime problems. Pareto charts, updated monthly, will help you spend your problem-resolution dollars effectively. (For a refresher on Pareto charts, see Chapter 14.) Cause-and-effect diagrams, like the one showed in Figure 18.1, help you organize your thinking. They are one statistical tool that many people in this country are borrowing from the Japanese. They were developed by Dr. Kaoru Ishikawa and are often called Ishikawa diagrams. They are also called fishbone diagrams because of their appearance.

Cause-and-effect diagrams identify variables contributing to an effect you want to control. In Figure 18.1, the effect is propellant burning rate for a solid propellant missile. The effect is shown at the head of the fisbone skeleton.

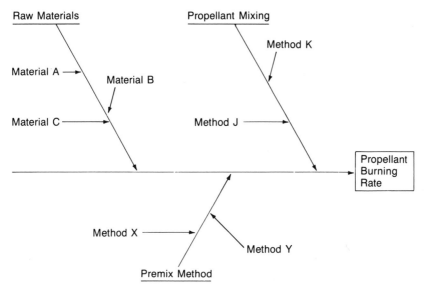

Figure 18.1. Cause-and-effect diagram of propellant processing.

Primary variables that might cause changes in the effect are identified as limbs attached to the trunk of the skeleton. Secondary variables sprout from the limbs representing the primary variables. Tertiary variables, if any, sprout from the smaller branches representing the secondary variables. This process continues until your diagram tells you all you want to know or becomes unreadable.

Figure 18.1 has been simplified.

Propellant burning rate is the effect being studied. It is at the head of the skeleton.

Raw materials, premix methods, and propellant mixing methods are the primary variables to be investigated. They are shown as limbs off the backbone or trunk.

Specific raw materials are smaller branches on the raw material limb. If particle size and other characteristics of the raw materials were included, they would become sprouts from their specific raw material branches.

Brainstorming sessions are often used when fishbone diagrams are being developed.

- The first step is to identify all primary variables that could contribute to the effect being studied (the limbs).
- Then identify sequences. In Figure 18.1, for example, raw materials come first. They are blended into a premix, using one of two possible premix procedures.
- The liquid premix ingredients are then mixed with the rest of the formulation using one of two possible mixing procedures.

Brainstormers can complete the diagrams and identify possible cause-and-effect relationships that influence propellant burning rates. Possible methods of identifying major contributors and screening unpromising variables are also brainstormed.

Cause-and-effect diagrams are similar to process flow charts. The main differences are in the form used and the detail presented. Few process flow charts go into as much detail as a fishbone chart.

Once the thought processes have been organized, it is time to start using more complex statistical tools.

When testing is inevitable divide-and-conquer techniques are an excellent way to keep costs down. (Details are shown in Chapter 26.)

Regression analysis is used to determine how one variable influences another. It helps predict what will happen to one variable when another changes. For example, "What happens to the viscosity of motor oil when the temperature changes?"

Although the calculations are complex, relatively low-priced computer programs are available to do the analyses for you. As an example, the "Number Cruncher Statistical System" by Dr. Jerry L. Hintze (865 East 400 North, Kaysville, Utah 84037) will run on most IBM-compatible home computers. It can do a variety of complex statistical calculations including regression analysis. It costs less than $100. (Details on regression analysis are shown in Chapter 27.) Many of the statistical tools mentioned in this chapter are complicated and require additional explanation. Most of them will be covered more thoroughly in subsequent chapters.

Although you may never go through detailed calculations like those required in regression analysis, you should be able to interpret the results. This includes understanding what the computer printouts mean and how the results can be used to solve production problems.

The executive who refuses to develop a basic understanding of statistical analyses and effective problem solving is like the manager who refuses to learn anything about computers. Both become captives of their technical staffs.

If managers know nothing about computers, their computer staff will advise them that:

- The voluminous computer reports they receive are exactly what they requested. When managers don't request summaries, they seldom get summaries.
- Changes in existing programs are impractical. This is a stock answer that means, "Sorry boss, but I really don't understand what you want."
- All computer problems can be solved by either buying bigger computers or raising the salaries of analysts and programmers.

If managers know nothing about statistical process control and effective problem solving, their technical staff may advise them that:

- All solvable problems have been solved.
- Scrap, rework, and liability suits are facts of life.
- The old way of doing things is the only way that works.
- People don't cause production problems; it's the gremlins and trolls.

As George Knucklehead once said, "I don't need to know the statistical side of the job. I can hire people to screw things up for me."

REFERENCES

Ishikawa, K. "What Is Total Quality Control?" Englewood Cliffs, N.J.: Prentice-Hall, 1985.

Hintze, J. L. "Number Cruncher Statistical System." Kaysville, UT: J. L. Hintze, 1985.

Choi, S. C. "Introductory Applied Statistics in Science." Englewood Cliffs, N.J.: Prentice-Hall, 1978.

Ott, E. R. "Process Quality Control." New York: McGraw-Hill, 1975.

Lindgren, B. W., and McElrath, G. W. "Introduction to Probability and Statistics." New York: Macmillan, 1966.

19

Basic Statistics

Statisticians are practical mathematicians and mathematicians are among the most controversial people in the world.

In **The Republic**, Plato (428-348 B.C.) wrote, "I have hardly ever known a mathematician who was capable of reasoning." Going to the other extreme, Havelock Ellis (1859-1939) claimed, "The mathematician has reached the highest rung on the ladder of human thought."

Most managers agree with Plato.

Mathematicians, however, developed statistics, and statistics are needed for statistical quality control, statistical process control, and statistical trouble-shooting. You don't have to be a statistician to use these tools, but you should understand the language. Managers who know nothing about statistics can be as dangerous as a compulsive spender with a dozen corporate credit cards.

MEANS AND MEDIANS

The most common statistical tools are means and medians. Both give an indication of how your data is centered. Unfortunately, they do not always give you the same answer.

The only "means" considered in this book are arithmetic means. They are calculated by adding the measurements and dividing the result by the number of observations.

A median is the middle point of the data. If you arrange all your observations according to size, the median will be the middle number.

Example 19.1

What are the mean and median of numbers 1, 2, 3? The mean is

(1 + 2 + 3) / 3 = 2.

- The sum of the numbers 1, 2, and 3 is 6.
- The number of observations is 3.
- Therefore, the mean is 6/3.
- The answer is 2.

The median is also 2 since it is in the middle when all the numbers are arranged according to size.

Example 19.2

What are the mean and median of numbers 0, 0, 1, 2, 2? The mean is

(0 + 0 + 1 + 2 + 2) / 5 = 1

- The sum of the numbers is 5.
- The number of observations is also 5.
- Therefore, the mean is 5 divided by 5.
- The answer is 1.

The median is also 1 since it is in the middle when all the numbers are arranged according to size.

With most data, the mean and the median will be approximately the same. Exceptions, however, occur when the data contains questionable observations called outliers. What is an outlier? An outlier is a measurement or observation that doesn't fit in with the rest of the data. It can be due to:

1. A gross error, like the transposition of numbers
2. A mechanical error, like having a computer keyboard that senses two keystrokes each time a key is touched
3. A mechanical breakdown, like a lathe with excessive chatter and worn-out bearings
4. A change in raw materials
5. An untrained operator or clerk
6. An attempt to fine tune a piece of equipment that was running smoothly

In other words, an outlier is data that doesn't fit.

The effect of an outlier is shown in the following example.

Example 19.3

What are the mean and median for numbers 1, 2, 2, 2, and 33? The mean is

$$(1 + 2 + 2 + 2 + 33) / 5 = 8.$$

- There are 5 observations and their sum is 40.
- Therefore, the mean is 40/5.
- The answer is 8.

The median is 2, since 2 is in the middle when the numbers are arranged according to size.

In this case, the mean and the median are quite different. The reason for the difference is the number 33, which looks out of place. It is an outlier and will normally be discarded. Most statisticians, however, investigate outliers before throwing them out.

Suppose the numbers in Example 19.3 represent error rates for five data entry clerks. If the fifth clerk was a trainee, 33 might be a valid number. Will supervision want to keep all of the records together? It is usually preferable to place trainee records in a separate file.

If the fifth clerk was not a trainee, what was the problem? Poor morale? Faulty equipment? Unproven computer software? Incorrect error rate calculations?

An investigation is warranted. Each type of problem requires different corrective action.

Means and medians will also differ from each other when the data is not normally distributed. Information on normal distribution is covered at the end of this chapter.

Once you know where the data is centered, you should determine how variable it is. Range, variance, and standard deviation are normally used for this purpose.

RANGE

Range is the difference between the highest number and the lowest number in the data.

Example 19.4

What is the range of numbers 1, 2, and 3?

Range = 3 − 1 = 2.

- 3 is the largest number and 1 is the smallest.
- The difference is 2.

Example 19.5

What is the range of numbers 5, 4, 4, 3, 3, and 2?

Range = 5 − 2 = 3.

- 5 is the largest number and 2 is the smallest.
- The difference is 3.

Example 19.6

What is the range of numbers 22.3, 17.5, 33.4, 66.8, 99.1, 9.5, and 33.4?

Range = 99.1 − 9.5 = 89.6.

- 99.1 is the largest number and 9.5 is the smallest.
- The difference is 89.6.

VARIANCE AND STANDARD DEVIATION

Statisticians normally refer to standard deviation as the Greek letter "sigma" and variance as "sigma squared." In this book, we refer to standard deviation as SD and variance as SD squared. SD squared is SD times SD. Why do we deviate from tradition? It is hard enough to master statistics without having to learn the Greek alphabet.

Variance equals the sum of the squares of (each individual value subtracted from the mean) and then divided by (the number of the observations minus one.)

Variance is included at this point in the book because it can be used to calculate the standard deviation (SD).

Standard deviation (SD) equals the square root of the variance.

Instead of going through laborious calculations, we assume the readers will have their variance and standard deviation calculations run by one of the following:

1. A statistician
2. A statistical calculator
3. A computer

Means and standard deviations do not always tell the same story. The following listing shows the means and standard deviations for data used in Examples 19.4, 19.5, and 19.6.

Example	Sample size	Range	Standard Deviation	Range/SD
19.4	3	2	1	2.0
19.5	6	3	1.05	2.9
19.6	7	89.6	31.7	2.8

The ratio of range to SD is not constant.

The standard deviations in the above example were calculated by a computer. Statisticians seldom do their calculations by hand. It is more important to know how to use the statistics than to know how to do the calculations.

Computer programs, however, should be checked by a statistician before they are used. Even commercial computer programs may have errors.

Table 19.1 and Figures 19.1 and 19.2 will help you understand how means and standard deviations are used in statistics and statistical trouble-shooting.

FREQUENCY DISTRIBUTING CHARTS

Suppose you had the following lengths:

40.1, 40.2, 40.2, 40.3, 40.3, 40.3, 40.4, 40.4, 40.4, 40.4, 40.4, 40.5, 40.5, 40.5, 40.5, 40.5, 40.5, 40.6, 40.6, 40.6, 40.6, 40.6, 40.7, 40.7, 40.7, 40.8, 40.8, 40.9

If you used a computer program to plot a frequency distribution chart or histogram, the chart might look similar to the one in Table 19.1.

The right side of Table 19.1 shows a frequency distribution chart made out of asterisks.

Table 19.1 Frequency Distribution Chart

Group	Low Point in Group	High Point in Group	Frequency in Group	Chart*
1	40.01	40.01 +	1	*
2	40.02	40.02 +	2	* *
3	40.03	40.03 +	3	* * *
4	40.04	40.04 +	5	* * * * *
5	40.05	40.05 +	6	* * * * * *
6	40.06	40.06 +	5	* * * * *
7	40.07	40.07 +	3	* * *
8	40.08	40.08 +	2	* *
9	40.09	40.09 +	1	*

For convenience, the data is arranged into nine groups. In this case, each group is .01 units wide; the high point of a group is essentially the same as its low point.

The frequency column identifies how many observations fit into each group. The 40.01 group has one observation. The 40.02 group was two observations. The 40.03 group has three observations. The 40.04 group has five observations, etc.

When frequency distribution charts are plotted by a computer, they are usually arranged with the frequencies plotted horizontally and the measurements plotted vertically. This procedure makes the chart a little more difficult to understand but the computer programming is easier.

When frequency distribution charts are plotted by hand, the frequencies are usually plotted vertically and the measurements are plotted horizontally. A typical example is shown in Figure 19.1.

In both instances, the charts approximate the shape of a bell. When this happens, both the curve and the distribution are called "normal." In other words, they are normal distributions and normal curves.

Figure 19-2 shows a frequency distribution chart where:

1. The data distribution is "normal."
2. The number of observations is very large.
3. The distance between points on the base of the curve is very small.

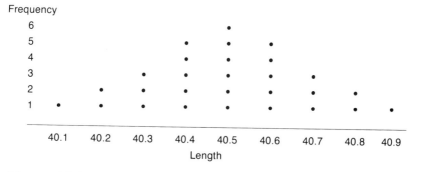

Figure 19.1 Frequency distribution chart.

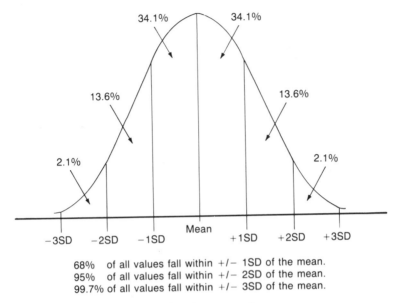

68% of all values fall within +/− 1SD of the mean.
95% of all values fall within +/− 2SD of the mean.
99.7% of all values fall within +/− 3SD of the mean.

Figure 19.2. Areas under the normal curve.

Frequency distribution charts meeting all of these criteria are called normal curves because they are so common. They are also called bell curves because they are shaped like bells. They are also Gaussian curves after the nineteenth-century mathematician who developed the fundamental laws of probability distribution.

Percentages shown in Figure 19.2 have been rounded off. When carried out further, they show that 68.3 percent of the data points will fall within ±1 SD of the mean, 95.4 percent of the data points will fall within ±2 SD of the mean, and 99.7 percent of the data points will fall within ±3 SD of the mean.

Each side of the mean will hold half of the data; normal curves are symmetrical.

If a point is more than three standard deviations away from the mean, it should be investigated. There are only three chances in a thousand that the point really belongs with the rest of the data. In view of these low odds, statisticians and engineers usually investigate data points that are different from the rest of the population.

These points are often considered outliers and are discarded. Then the standard deviation is recalculated.

Data used in Example 19.3 will help illustrate the effect of outliers on the standard deviation in this next example.

Example 19.7

Data with outlier: 1, 2, 2, 2, 33
Data without outlier: 1, 2, 2, 2
Statistics:

	With Outlier	Without Outlier
Mean	8	1.75
Median	2	2
Range	32	1
Standard deviation	14	.5

How do you use this information? Example 19.8 should help answer this question.

Example 19.8

You are working for the U.S. Bureau of the Census and have 500 technicians making statistical calculations for your directorate. All have essentially the same volume of production.

The average error rate of the group is 5.0 percent and the standard deviation is 1.0 percent. A technician transferred to your organization has an error rate of 6.0 percent. Would you treat this technician's error rate like an outlier? Assume the distribution is perfectly normal.

- Given:
  Mean   = 5.0
  Standard deviation = 1.0
  Suspect data   = 6.0

- Solution:
  Deviation from the mean = suspect data − mean
  $$= 6.0 - 5.0$$
  $$= 1.0$$

$$\frac{\text{Deviation from the mean}}{\text{Standard deviation}} = \frac{1.0}{1.0} = 1.0$$

In other words, the new technician's error rate is 1.0 percent greater than the mean. The standard deviation is also 1.0 percent. Therefore, the technician's error rate is 1 SD greater than the mean.

Since the questionable data is less than three standard deviations from the mean, it is not an outlier.

How does the new tech compare? Since the distribution of the data is perfectly normal, 50.0 percent of the techs can be expected to have error rates below the mean; they would have done better than the new tech.

Based on Figure 19.2, an additional 34.1 percent had error rates between the mean and 1 SD above the mean; they should have outperformed the new tech. Therefore, over 84 percent of the old technicians should have done better than the new one (50.0 + 34.1 = 84.1). On the other hand, 15.9 percent should have done worse (100.0 − 84.1 = 15.9). The new transfer was pretty good for someone new to the job.

Statistics may be painful but they do help solve problems.

SUMMARY

Means, medians, ranges, standard deviations, and frequency distributions are basic tools of the statistician.

Means, medians, and ranges are easy to calculate. Standard deviations are more complex, especially when used with large quantities of data. Their meaning, use, advantages, and shortcomings, however, are more important than the calculations.

To ease the drudgery, we use computers to do most of the math. Computers, with their speedy computations and functional graphics, are great as long as you understand their output.

There are few things more useless, however, than volumes of automated statistical analyses that no one understands.

20

Process Capability

Few processes are perfect. How good is yours?

Anyone who can't answer this question needs to learn about process capability studies.

Process capability studies show you the capability of your process --under ideal conditions.

In statistical terms, process capability range is equal to ± 3 SD or 6 SD in all.

Example 20.1

What is the process capability range for making cans to the proper height? The standard deviation is .01 in. according to a recent process capability study.

$$\text{Process capability range} = \pm 3 \text{ SD}$$
$$= 6 \text{ SD}$$
$$= 6 \times .01$$
$$= .06 \text{ in.}$$

Example 20.2

What is the process capability range for checking travel vouchers if the standard deviation is .2 errors per voucher? The standard deviation was determined by a process capability study.

Process capability range $= 6$ SD
$= 6 \times .2$
$= 1.2$ errors per voucher

PROCESS CAPABILITY VERSUS SPECIFICATIONS

Under ideal conditions, your discrepancies will be 0.3 percent if:

• Your specification is 6 SD wide.
• Your process average is in the middle of the specification.
• Your data is normally distributed.

Based on this information, you might be tempted to set your specification range at 6 SD. Don't!

If your specification range matches your process capability range, your discrepancy rate will exceed 0.3 percent because capability studies are run under ideal conditions.

Figure 20.1 shows what to expect when your specification range is the same as your process capability. Figure 20.1, like Figure 19.2, assumes that your data have a normal distribution.

No one can make the process average hit the middle of the specification all the time. Under the ideal conditions of a capability study you may come close. In production, you are lucky if your process average stays within 1 SD of the middle of the specification.

All examples in Figure 20.1 have the specification range equal to the process capability range. In addition, there are many measurements and their frequency distribution is "normal" (shaped like a bell curve).

The top curve in Figure 20.1 shows what happens when the process average is at the middle of the specification. Under these conditions, .13 percent of the production will be below the lower specification limit. Another .13 percent of the production will be above the upper limit.

The middle curve in Figure 20.1 shows what happens when the process average is 1 SD below the middle of the specification. Under these conditions, 2.3 percent of the production will be below the lower specification limit. Practically no production will be above the upper specification limit.

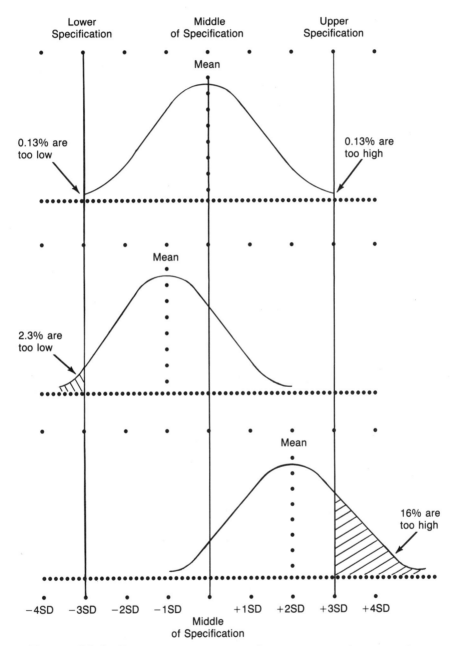

Figure 20.1 Process mean versus discrepancies when specification range is 6 SD.

The lower curve in Figure 20.1 shows what happens when the process average is 2 SD above the middle of the specification. Under these conditions, 16 percent of the production will be above the upper specification limit. Practically no production will be below the lower specification limit.

Figure 20.2 reflects the variability of processes found in production. All examples in Figure 20.2 have the specification range set at 4 SD.

Why is the specification so tight?

Suppose a process capability study showed that the standard deviation was 2.0 inches under ideal conditions. Then design personnel set the specification range at 6 SD. Based on these data, the specification would be 12.0 inches wide (2.0 inches per SD × 6 SD = 12.0 inches).

Suppose the standard deviation of the process increased to 3.0 inches due to variations in raw materials, equipment, and operators. Based on production data, the same 12-unit specification would be only 4 SD wide (12 inches / 3.0 inches per SD = 4 SD).

All curves in Figure 20.2 have a specification that is 4 SD wide. In addition, there are many measurements and their distribution is normal.

In the top curve, the process average falls on the middle of the specification. Under these conditions, the expected discrepancy rate is 2.3 percent on the low side of the specification and 2.3 percent on the high side.

The middle curve in Figure 20.2 shows what happens when the process average is 1 SD below the middle of the specification. In this case, the expected discrepancy rate is 16 percent on the low side and .13 percent on the high side.

The bottom curve in Figure 20.2 shows what happens when the process average is 2 SD above the middle of the specification. In this case, there are practically no defects below the lower specification limit. Half of the production is above the upper specification limit.

For a given standard deviation, the closer the process average is to the middle of the specification, the fewer the discrepancies. On the other hand, defects increase when the process average shifts away from the middle of the specification.

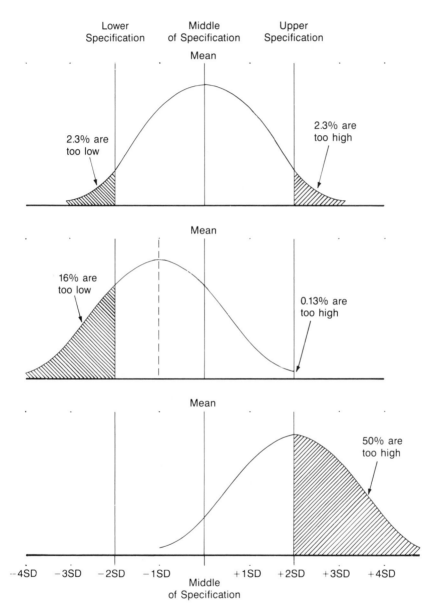

Figure 20.2 Process mean versus discrepancies when specification range is 4 SD.

The relationship between the process average, the specification limit, and the percent discrepancies is shown by the following table.

Table 20.1 Percent Out of Specification Versus Distance of Process Average from Specification

Distance of process average from specification limit (in standard deviations)	Relationship to specification	
	Out of spec	In spec
0.0 SD	50.0%	50.0%
0.0 SD	46.0%	54.0%
0.2 SD	42.1%	57.9%
0.3 SD	38.2%	61.8%
0.4 SD	34.5%	65.5%
0.5 SD	30.8%	69.2%
0.6 SD	27.4%	72.6%
0.7 SD	24.2%	75.8%
0.8 SD	21.2%	78.8%
0.9 SD	18.4%	81.6%
1.0 SD	15.9%	84.1%
1.1 SD	13.6%	86.4%
1.2 SD	11.5%	88.5%
1.3 SD	9.7%	90.3%
1.4 SD	8.1%	91.9%
1.5 SD	6.7%	93.3%
1.6 SD	5.5%	94.5%
1.7 SD	4.5%	95.5%
1.8 SD	3.6%	96.4%
1.9 SD	2.9%	97.1%
2.0 SD	2.3%	97.7%
2.2 SD	1.4%	98.6%
2.4 SD	0.8%	99.2%
2.6 SD	0.5%	99.5%
2.8 SD	0.3%	99.7%
3.0 SD	0.1%	99.9%

Note: SD = standard deviation and spec = specification. Assumptions include: (1) The distribution is normal; and (2) The process average or mean is between the lower and upper specifications limits.

The following example will help you apply the data shown in Table 20.1.

Example 20.3

- The lower specification limit is 40.0.
- The upper specification limit is 42.0.
- The process average (mean) is 40.5.
- The standard deviation of the data is .3.
- What percent of production will be outside the specification if nothing is changed?

$$\text{Spec range in SD} = \text{(spec range)/SD}$$
$$= (42.0 - 40.0)/0.3$$
$$= 6.7 \text{ SD}$$

- Low side:

$$(\text{mean} - \text{spec.})/SD = (40.5 - 40)/.3$$
$$= 1.7 \text{ SD}$$

Table 20.1 specifies 4.5 percent defectives when the process mean is 1.7 SD above the lower specification limit.

- High side:

$$(\text{spec.} - \text{mean})/SD = (42.0 - 40.5)/0.3$$
$$= 5 \text{ SD}$$

Table 20.1 specifies .1 percent defectives when the process mean is 3 SD below the upper specification limit. The percentage of defectives is much smaller when the distance is 5 SD.

- Total: The total percentage of defectives is the percentage of defectives on the low side added to those on the high side. In this case it is 4.5 + .0, or 4.5 percent defectives

HOW TO RUN A PROCESS CAPABILITY STUDY

When running a processs capability study you should:

1. Ensure that everything is in good condition.
2. Select a competent operator to run the equipment.
3. Control critical variables.
4. Produce at least 30 good, consecutive parts.
5. Record data in production sequence.
6. Calculate the process capability.

Use manufacturing equipment that is in good condition. Don't select a lathe with sloppy bearings or a keypunch machine with sticky keys.

Choose an operator who is representative of the production crew. Trainees and problem operators should not be selected.

Keep tight controls on critical variables. With a lathe, for example, control the raw material, coolant, spindle speed, and feed rates.

Make adjustments until the process average is in the center of the specification. Then produce at least 30 good, consecutive parts.

Avoid making adjustments, if possible. Otherwise, record the number of the first part following the adjustment, then make three extra parts at the end of the run.

Measure the product and record the results. The data should be arranged in production sequence and shown in groups of three consecutive parts. Discard any group of three containing a part that immediately followed a machine adjustment.

Calculate the range and average for each group of three. Then calculate the overall average and the average range for all of the groups.

Multiply the average range by 3.544. The result is your process capability.

A typical example is shown in Table 20.2.

The formula: "Process capability = average range × 3.544" assumes there are no outliers or wild points. It also assumes the distribution is normal and the ranges are calculated from groups

Table 20.2

Sequence of group (3 parts per group)	Sequence in group			Range of group	Average of group
	1	2	3		
1	20	21	22	2	21.0
2	21	20	22	2	21.0
3	20	21	21	1	20.7
4	22	20	20	2	20.7
5	20	22	19	3	20.3
6	20	22	20	2	20.7
7	20	21	20	1	20.3
8	21	20	19	2	20.0
9	20	19	21	2	20.0
10	19	22	20	3	20.3
Total				20	205.0
Average				2.0	20.5

Process capability = average range × 3.544

= 2.0 × 3.544

= 7.1

Process capability = 6 standard deviations

Therefore, one standard deviation = 7.1/6

= 1.2

of three data points. Factors other than 3.544 are used when the group size is not three.

Once the calculations are complete, you can either select your specifications or verify that existing specifications are reasonable.

Where product performance permits, the specification range should exceed 8 SD. Why?

- Capability studies are based on ideal conditions. Production operations are not.
- Specifications should be based on real needs, not impossible dreams.

- Specifications that can't be met won't be.
- Frustrated inspectors, however, have been known to accept discrepant parts when nothing else is available.
- If specifications are violated and there is no penalty, they will be violated again.
- If specifications are tighter than necessary, they will increase costs without increasing value.

CRITICAL QUESTIONS

Questions that should be answered by design engineers before they set unrealistic specifications include:

1. Will raw materials used by production be the same as those used during preproduction testing?
2. Do all similar machines have the same precision and accuracy?
3. Are all production operators as skilled as those used in the process capability studies?
4. Can production personnel consistently operate within the process capability limits?

The answer to each of these questions is no.

Every delivery of raw materials will be different from the previous one. Sometimes differences are small. Sometimes they are large enough to hurt production, even though all deliveries pass receiving inspection. Product specifications must be broad enough to allow for these variations.

Similar machines may have similar precision, but some differences are inevitable. As machines and tooling wear, these differences increase. Engineering personnel should not assume that every machine in the shop can meet the performance achieved during the process capability tests.

The accuracy and precision of different operators also varies. Specifications must reflect this reality.

Since process capability tests are run under ideal conditions, the specification range should exceed the process capability by at least 30 percent. This allows for variations in raw materials, machines, and operators.

If tighter specifications are necessary:

- Tighten the raw material specifications.
- Improve the process.
- Get better equipment.
- Give the operators additional training.
- Change the design.

Making specifications tighter than the process capability is like wishing on a star. As John Ray (1627-1705) wrote in **English Proverbs**, "If wishes were horses, beggars might ride."

Summary

Process capability studies use most of the statistics covered so far.

If you don't know the capability of your process, you are apt to set specifications that are unrealistic. Specifications that can't be met are like gigantic fortresses made of papier-mâché: They don't provide much protection.

If you don't have some knowledge of means and standard deviations, you will have a hard time determining your process capability.

21

Which Product Is Best?

Some of the worst quality disasters occur when someone wants to make a minor change like substituting a new raw material or redesigning a functional product. Unfortunately, few people give adequate thought to the question, "Is the new better than the old?" One of the easiest ways of exploring this question is to run an ends test.

Ends testing was developed by Dorian Shainin and his associates when they were working at Rath & Strong, Inc., a prominent management consulting firm. The tool is simple. The calculations are even simpler.

In its basic form, ends testing involves running a series of performance tests on two products or processes you want to compare. Then you list them in the order of the test results. Normally, products with the best results are shown on top.

Let's call the items or processes A and B. If the worst results for item A look better than the best results for item B, then item A is the winner. Confidence in the results depends on the sample sizes. Data on confidence levels and sample sizes is shown in Table 21.1.

The basic procedures include:

1. Defining the test conditions
2. Determining how important it is to be right
3. Selecting the sample sizes
4. Selecting the testing sequences
5. Determining how much improvement you want

Table 21.1   Criteria for Evaluating Ends Tests

Confidence level for decision	Sample size for first material	Sample size for second material
99%	2	13
	3	7
	4	5
95%	2	5
	3	3
	5	2
90%	1	9
	2	3
	3	2

Test conditions must be rigidly controlled. Existing materials should be tested using regular procedures, unless procedures are being evaluated. New materials should be tested using conditions you plan to impose if the new material is accepted.

Conclusions are valid for the procedures used in the test. They should not be extrapolated to conditions that haven't been evaluated.

Be fair. If, for example, you are testing tires for use with company cars, make sure the test conditions are equitable. You don't want one driver burning rubber throughout the test and another driver acting like there were twelve dozen eggs on the back seat. Under those conditions, the test results would not be valid.

How important are the results?

If being right is critical, use the 99 percent confidence level. Ninety-nine percent confidence means that there is only one chance in a hundred that you will make an incorrect decision. For 99 percent confidence, select sample sizes from the top three choices in Table 21.1.

If an erroneous decision is easily corrected, select one of the last three choices; you will save money. Tests using the bottom set of sample sizes have a 90 percent chance of giving you the right decision. In other words, there is one chance in ten that a product will be favored when it shouldn't be.

The range of sample sizes in Table 21.1 was designed to reflect differences in the availability or cost of competing products and processes. The table gives more options than it seems since either product can be identified as the "first material."

When using Table 21.1, the "first material" might represent a competitor's product that is hard to get. The "second material" might represent your product that is readily obtainable. In this instance, you would prefer to limit the use of the competitor's product to one or two tests.

When the cost and availability of the items is similar, use approximately the same number of tests for each candidate. When working at a 99 percent confidence level, for instance, you would probably prefer to test 4 of the competitor items and 5 of your own. The total number of tests is 9. If you used 2 of the competitor's items and 13 of your own, the total number of tests would be 15.

A random test sequence is required for ends testing. There is always the possibility of a time related variable influencing the results, especially when you test all of one type of product before starting on the second.

The sequence of random testing can be determined by:

- Using a table of random numbers
- Drawing numbers out of a hat
- Using a computer-generated test sequence
- Using any other method that ensures each item has an equal chance of being selected for each test sequence.

How much improvement do you want?

If you were testing two brands of tires for mileage, the price of the tire might be a factor. As an example, suppose one tire cost 50 percent more than the other. You would expect the more expensive tire to give you 50 percent more mileage in order to be competitive.

Example 21.1

- You are testing abrasion resistance of two new inks.
- You want 99 percent confidence that you made the right choice.
- The inks have about the same price and availability.

Which test plan would you use?

Solution 21.1

If you made four runs with the first ink and five runs with the second, you would have a total of nine test runs. Since both inks are readily available and cost of the inks is not a factor, you would use this option. Among the sample plans that provide 99 percent confidence, it requires the least testing.

Example 21.2

The results of the test in example 21.1 were as follows:

Ink	Test sequence	Number of rubs needed to remove ink
A	4	50
A	1	48
A	5	47
A	9	45
B	8	40
B	2	39
B	6	38
B	3	38
B	7	37

Which ink would you choose?

Solution 21.2

In this case, there is a five-unit gap between the worst sample of ink A and the best sample of ink B. There is no overlap and and there is no reason for favoring ink B due to price or other special factors.

Ink A passes the ends test. You have a 99 percent confidence that your choice of inks was justified.

Note the sequencing of the tests. The first test used ink A. The second used ink B. The third used ink B. The fourth used ink A . . . The sequence was random.

Example 21.3

- You are testing catalysts for a chemical plant. Catalyst A costs more than catalyst B. Because of this price difference, your engineers recommend catalyst B unless catalyst A's yield is better by at least 5 percent (absolute).
- You think 10 to 1 odds are pretty good, so you will accept a 90 percent confidence level.
- The tests are much more expensive than the catalyst, so you select the qualifying plan with the least number of tests.
- Catalyst A gets the fewest tests because it costs more.
- The test results are:

Catalyst	Test sequence	Product yield
A	4	94.0
A	2	93.2
B	1	88.0
B	3	87.0
B	5	84.5

Which catalyst would you choose?

Solution 21.3

Before the test you determined that catalyst A wouldn't be selected unless it had a yield 5 percent better than catalyst B. Therefore, it is necessary to add 5 percent to each catalyst B result. Then, you reevaluate the ranking according to yield. The new rankings are:

Catalyst	Test sequence	Product yield (adjusted)
A	4	94.0
A	2	93.2
B	1	93.0 (88.0 + 5.0)
B	3	92.0 (87.0 + 5.0)
B	5	89.5 (84.5 + 5.0)

In this case the rankings didn't change. The yields are closer but the best run using catalyst B is still worse than the poorest run using catalyst A, even after the adjustments. Use catalyst A.

Example 21.4

- You are comparing two computer programs. The risk of making the wrong decision is relatively low so you accept a 90 percent confidence level.
- You choose to make two tests using the first program and three tests using the second program.
- Speed of data entry is your method of evaluation.
- The results are:

Program	Test Sequence	Data entries per shift
A	1	511
B	5	501
A	4	490
B	3	440
B	2	435

Which program should you choose?

Solution 21.4

In this example, there is no clear-cut decision. Neither product is better all the time; the test results overlap.

You might decide to use program A rather than pay for more testing; the average results using program A are a little better. You would not, however, have a 90 percent confidence in your decision.

An alternative is to drop ends testing and use a more classical method of analysis. Ends testing is simple and effective but it has limitations.

When more detailed information is needed and cost is not critical, more formal analyses can be used. They will be covered in subsequent chapters.

SUMMARY

Ends tests help you make objective decisions without an excessive commitment of time and money. Advantages of ends tests include:

1. They are relatively inexpensive.
2. They help you make objective decisions.
3. They are easy to interpret.

Disadvantages of ends tests include:

1. They only apply to the conditions of the test.
2. They do not work well when differences between processes or materials are small.

The next time you consider switching materials or processes based on a gut feeling, consider verifying your decision with ends tests. They may help keep the gut feeling from turning into a stomachache.

22

Is the New Way Really Better?

Ends tests are an excellent way of determining whether a new process or product is better than the old. There are times, however, when ends tests do not give you a clear-cut decision. Each problem has its own requirements that must be considered in an analysis.

The available methods of analysis are as diverse as the statisticians who invented them. Some techniques require assumptions and mathematical manipulations that can give you migraines. Most of these techniques will be left for college textbooks on statistics. We are not trying to make everyone into statisticians; we are merely familiarizing you with some technical terms and tools that statisticians use. It is much easier to keep from being snowed when you know a little about the language. Statisticians love to snow novices.

Additional techniques for comparing processes and materials without excessive trauma include:

- The Tukey-Duckworth test
- The paired t-test

The names of these analyses are more formidable than the tests themselves, and neither one is limited to data that follows a "normal" distribution. They don't even require that the two sets of data have the same standard deviation; some formal tests do.

Relieved?

THE TUKEY-DUCKWORTH METHOD

The Tukey-Duckworth method was published in the February 1959 issue of **Technometrics**. Steps in the method include:

1. Counting the number of measurements of material A that exceed the highest values of material B, assuming material A has the largest values.
2. Counting the number of measurements of material "B" that are below the lowest value of material "A". If this value is zero, the test concludes there is no significant difference between the materials.
3. Add the results of steps 1 and 2.

When using the Tukey-Duckworth method, the approximate risk of making a bad decision is shown in Table 22.1.

Example 22.1

In example 21.4, we were not able to establish a risk level when trying to decide between two computer programs. Can we do better with the Tukey-Duckworth method?

Table 22.1 Critical Sums Using the Tukey-Duckworth Method

Aproximate risk (%)	Count at step 3
9	6
5	7
1	10
0.1	13

The data were as follows:

Program	Data entries per shift
A	511
B	501
A	490
B	440
B	435

Solution 22.1

- Step 1: There is one A above the highest B.
- Step 2: There are two B's below the lowest A.
- Step 3: The sum of the results of steps 1 and 2 is 3. Therefore, the risk of choosing program A when program B should have been selected is greater than 9 percent (see Table 22.1).

Actually, the Tukey-Duckworth test is not appropriate for evaluating fewer than six observations. Under ideal conditions, six observations can identify risks as low as 9 percent. A minimum of ten observations are needed if you want to know whether the risk is below 1 percent.

Example 22.2

Using the data that follow, what is the risk of choosing the wrong delivery service? The decision is to be based on promptness of delivery. All deliveries are between the office and a station 100 miles away.

Delivery sequence	Service	Hours to make the delivery
1	B	4.8
2	A	5.0
3	B	6.3
4	B	5.1
5	A	3.0
6	A	4.2
7	B	7.0
8	A	3.3
9	B	6.2
10	A	3.1

Solution 22.2

It is easier to evaluate the delivery services if data are listed according to performance.

Delivery sequence	Service	Hours to make the delivery
5	A	3.0
10	A	3.1
8	A	3.3
6	A	4.2
1	B	4.8
2	A	5.0
4	B	5.1
9	B	6.2
3	B	6.3
7	B	7.0

- Step 1: There are 4 A's above the highest B.
- Step 2: There are 4 B's below the lowest A.
- Step 3: 4 + 4 = 8.

Table 22.1 shows that the approximate risk of a wrong decision when step 3 adds up to 8 is between 1 and 5 percent. It is traditional to take the conservative number; call the risk 5 percent.

So far we have investigated the easy way of comparing processes or materials. Advantages of these techniques include:

- They involve little mathematics.
- They do not assume "normal" populations.
- They do not assume equal standard deviations for the two sets of data.

Statistical analyses that meet these free and easy criteria are called "nonparametric."

Any time you go beyond nonparametric analyses of whether two processes or products are different, it is wise to consult a statistician. Otherwise, you may be making assumptions that are not justified.

One of the procedures used by statisticians to compare processes and products is the paired t-test.

PAIRED T-TESTS

The paired t-test is "nonparametric," but it still requires a painful amount of math. Computer programs, however, are available to reduce the frustrations of those who use this analysis.

The paired t-test is used when the data come in pairs—one from the first material or process; one from the second. The size of the samples from each material or process must be the same.

As an example, suppose you wanted to know which of two computer programs was the fastest. You could use a single computer to run a problem twice—once with the first program and once with the second. This would give you one set of paired data. Then you could introduce additional problems and pair results. Each computer problem would give you two completion times, one for each computer program. A typical use of paired data follows:

Example 22.3

The following data were obtained from testing computer programs A and B using five different problems. A single computer was used during the test. The testing sequence was alternated. With problem 1, program A was used first. With problem 2, program B was used first.

Problem	Minutes required to solve problem	
	Using program A	Using program B
1	3.0	4.8
2	6.1	6.2
3	2.1	3.0
4	7.5	7.4
5	4.0	5.0

Which computer program was fastest?

Solution 22.3

These data were analyzed using Dr. Hintze's NCSS computer program. The results indicate there was a 95 percent probability that program A was the fastest under the conditions of the test.

Neither the ends test nor the Tukey-Duckworth method would be effective in analyzing this data. Why not?

There was too much variability in the problems that were used in the evaluation. The wide range of problems, however, was necessary. You want to give the programs a fair test. One program might be best with one type of problem. The other program might be best with another.

Even in this example, there is some doubt about which program is fastest at solving the more complex problems. As an example, the one time that program B beat out program A was when the execution time was above seven minutes; the other runs took less time.

Example 22.4

What results would you get from the data in example 22.3 using the ends test or the Tukey-Duckworth test?

Solution 22.4

Both the ends test and the Tukey-Duckworth test require arranging the data according to the time taken to solve the sample problems. The resulting rearrangement shows:

Computer program	Minutes taken to solve problem
A	2.1
B	3.0
A	3.0
A	4.0
B	4.8
B	5.0
A	6.1
B	6.2
B	7.4
A	7.5

Using the ends test as presented in this book, there is too much overlap to identify the risk involved in selecting program A.

The Tukey-Duckworth test is also unable to identify the risk.

- Step 1 of the Tukey-Duckworth test shows only one instance where program A did better than the best run using program B.
- Step 2 shows no instance where program B was slower than the slowest time of program A.
- When step 2 is zero, the Tukey-Duckworth test is unable to determine the risk of accepting one of the programs.

SUMMARY

When comparing materials or processes, the ends test, the Tukey-Duckworth test and the paired t-test each has advantages and disadvantages.

The ends test generally requires the fewest observations. The test is easy to run and requires few calculations.

The Tukey-Duckworth test generally requires more observations than the ends test; its greatest advantage comes when results overlap. Like the ends test, it is easy to run and requires few calculations.

The paired t-test is good when you intentionally change variables as part of the test. It is easy to run but requires fairly difficult math. The trauma of solving complex equations, however, is no longer a serious drawback. Computers can do the calculations for you.

23

Frequency Distribution Charts

An old proverb is, "One picture is worth more than a thousand words." Frequency distribution charts are just as useful.

Frequency distribution charts, alias histograms, were discussed in Chapter 19, under "Basic Statistics" and in Chapter 20 under "Process Capability." The one question these chapters addressed was, "Is the data normally distributed?" When the measurements are not normally distributed, we should find out why.

Figure 23.1 shows a frequency distribution chart for the inside diameter of a hypothetical batch of close tolerance metal sleeves.

Information that can be obtained from this chart includes:

- The distribution is normal; the curve is shaped like a bell.
- All units fall within the specification limits; the specification is compatible with the process capability.
- The median is in the middle of the specification; production deserves credit for keeping the process on target.
- There is a 0.001-inch safety margin between the lowest unit and the lower specification limit.
- There is also a 0.001-inch safety margin between the highest measurement and the upper specification limit.

Frequency distribution charts, like the one in Figure 23.1, show the process is in control and the specification is compatible with the process capability.

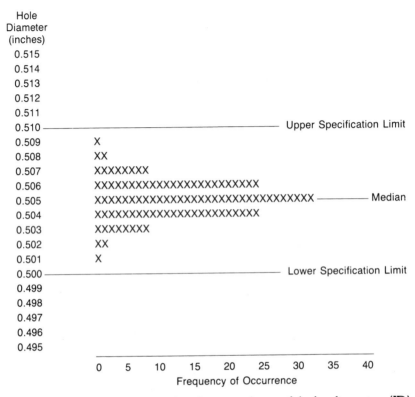

Figure 23.1 Frequency distribution chart of hole diameter (ID) showing normal distribution and good process control.

Figure 23.2 is a frequency distribution chart for the same part shown in Figure 23.1. This time, however, the operation is having problems.

Information that can be obtained from this chart includes:

- The distribution is normal; the curve is shaped like a bell.
- The median is in the middle of the specification; production personnel deserve credit for keeping the process on target.
- Production is out of specification on the low side as well as the high; the specification range is too tight for the process capability.

The process capability should be improved. If this is not possible, the specification should be relaxed. The alternative to these suggestions is extensive scrap and rework. This much scrap and rework should not be tolerated if other acceptable alternatives are available.

When the specification and the process capability are not compatible:

1. The specification should be reviewed to determine whether the limits can be relaxed without hurting either the quality or the safety of the product. Some specifications are based on need. Others are based on whim.

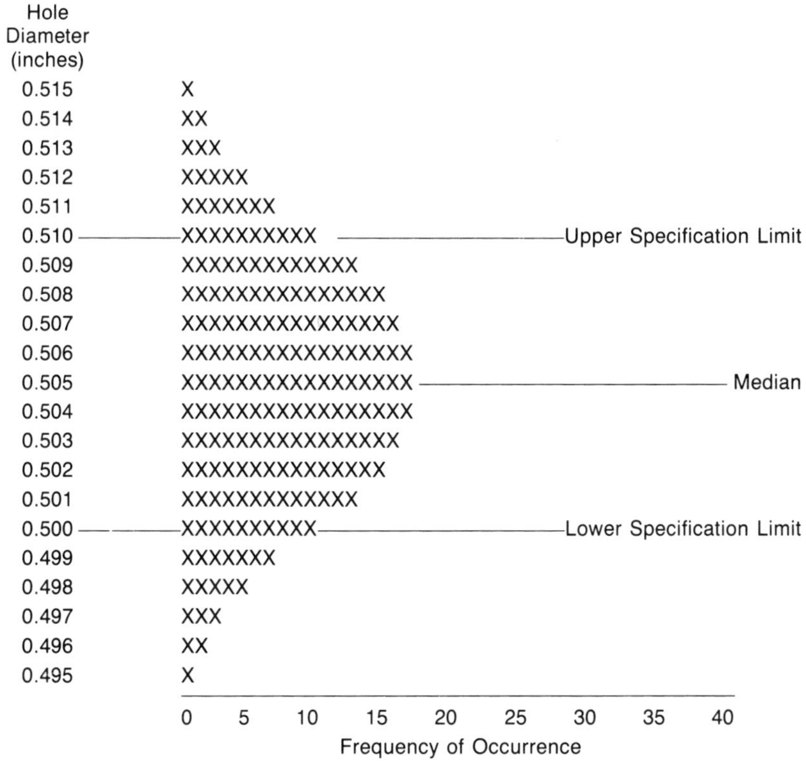

```
Hole
Diameter
(inches)
 0.515        X
 0.514        XX
 0.513        XXX
 0.512        XXXX
 0.511        XXXXXX
 0.510 ———————XXXXXXXXXX  ———————————————————Upper Specification Limit
 0.509        XXXXXXXXXXXXX
 0.508        XXXXXXXXXXXXXXX
 0.507        XXXXXXXXXXXXXXXX
 0.506        XXXXXXXXXXXXXXXXX
 0.505        XXXXXXXXXXXXXXXXX ———————————————————— Median
 0.504        XXXXXXXXXXXXXXXX
 0.503        XXXXXXXXXXXXXXX
 0.502        XXXXXXXXXXXXXX
 0.501        XXXXXXXXXXXX
 0.500 ———————XXXXXXXXXX————————————————————Lower Specification Limit
 0.499        XXXXXXX
 0.498        XXXXX
 0.497        XXX
 0.496        XX
 0.495        X
            0    5   10   15   20   25   30   35   40
                      Frequency of Occurrence
```

Figure 23.2 Frequency distribution chart of hole diameter (ID) showing a normal distribution and an incompatibility between the process capability and the specification.

2. The process should be reviewed. Are better processes, procedures, equipment, or materials available?
3. Specialized vendors should be considered. In some instances, the cost of improving your equipment is excessive; but local vendors may already have the skills and equipment that are needed.

Figure 23.3 represents a frequency distribution chart for a process that should be free of defects.

Information that can be obtained from this chart includes:

• The data is normally distributed; the curve is shaped like a bell. In this case, the bell is very narrow; this chart shows little variability compared to Figures 23.1 and 23.2.

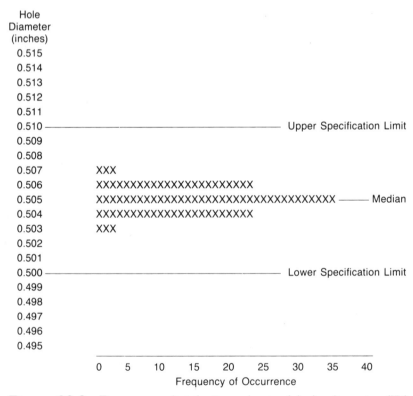

Figure 23.3 Frequency distribution chart of hole diameter (ID) showing normal distribution and excellent process control.

- The median is in the middle of the specification. As with the other charts, production deserves credit for keeping the process on target.
- All data are within specification limits that are compatible with the process capability.
- If this chart represents parts received from a supplier, it may not be necessary to pass every lot through receiving inspection. Special techniques such as continuous sampling, skip lot sampling, or a ship-to-stock program might be considered. Details on these techniques should be obtained from vendor quality control specialists.

Figure 23.4 represents a frequency distribution that is essentially flat.

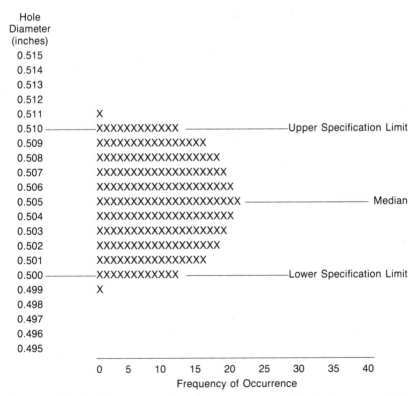

Figure 23.4 Frequency distribution chart of hole diameter (ID) showing a truncated distribution.

Conclusions that might be drawn from this chart include:

- The data distribution is **not** normal; it is truncated. A truncated distribution has one or both ends cut off.
- The median of the data is in the middle of the specification; the manufacturer deserves credit for keeping the process on target.
- The specification was too tight for the process capability of the producer. The distribution may have been normal when the parts were made but it was changed.

How?

When a producer is not able to control the process to meet the specification, one option is to inspect every part. Parts that meet the specification are shipped. Those that don't meet the specification were either reworked, scrapped, or sold to companies with looser specifications.

With 100 percent inspection, some questionable parts usually slip through; there is always some difference between the vendor's inspection and receiving inspection.

Figure 23.5 approximates the bimodal distribution found in a West Coast plant making filament-wound chambers.

The term "bimodal" refers to the two peaks or modes shown on the chart. One peak or mode occurred between .021 and .025 inches. Another peak or mode occurred between .051 and .055 inches.

Because of the wide range of chamber shrinkage patterns, the company had extensive scrap and rework of their chamber skirts.

The chamber length specification was only .040 inches wide. When a chamber was machined with the assumption that it would shrink .022 inches and it actually shrank .053 inches, it usually ended up being too short.

The bimodal population in the frequency distribution chart tipped off the engineers. They were convinced that different processes or materials caused the two peaks. In this case, there were two processes.

The shop had developed a quick, high-temperature cure cycle that they used when schedules were tight. Filament-wound chambers that were cured quickly had a different shrinkage pattern than those cured slowly.

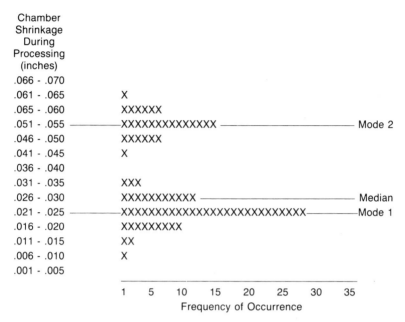

Figure 23.5 Frequency distribution chart of chamber shrinkage showing bimodal distribution.

Use of the frequency distribution chart to identify the two populations reduced the time required to find the problem. Without this information, the engineers might have concentrated on tightening the process controls.

In many instances, the process capability of either population fits within the specification range. The apparent process capability of the combined populations, however, is too broad.

Bimodal frequency distribution charts are seldom as clear cut as the one shown in Figure 23.5. In most cases there is extensive overlap between the two chart populations.

Figure 23.6 shows a common example—the heights of college students. The heights of male and female students are included in the same chart.

The chart shows two peaks or modes. The peak that occurs at (62-63 inches) would normally have more women than men. The

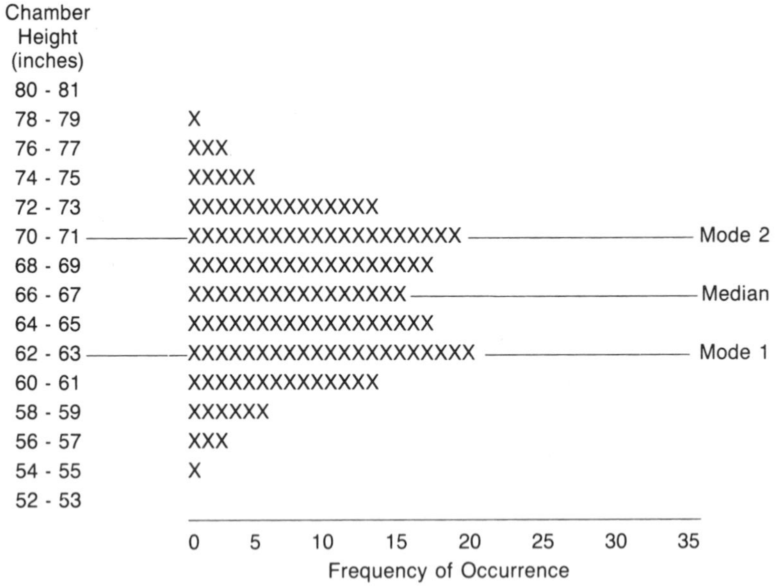

```
Chamber
Height
(inches)
 80 - 81
 78 - 79        X
 76 - 77        XXX
 74 - 75        XXXXX
 72 - 73        XXXXXXXXXXXXX
 70 - 71 ───────XXXXXXXXXXXXXXXXXXXX ─────────────── Mode 2
 68 - 69        XXXXXXXXXXXXXXXXX
 66 - 67        XXXXXXXXXXXXXXX ──────────────────── Median
 64 - 65        XXXXXXXXXXXXXXXXX
 62 - 63 ───────XXXXXXXXXXXXXXXXXXXXX ────────────── Mode 1
 60 - 61        XXXXXXXXXXXXX
 58 - 59        XXXXXX
 56 - 57        XXX
 54 - 55        X
 52 - 53

               0    5    10   15   20   25   30   35
                       Frequency of Occurrence
```

Figure 23.6 Frequency distribution chart of heights of college students showing bimodal distribution.

peak that occurs at (70-71 inches) would normally contain more men than women. You can, however, expect some overlap. Some men usually fall in the (62-63 inch) group. Some women usually fall into the (70-71 inch) group.

If you had the frequency distribution chart, with no indication of whether it included both men and women, the chart itself would give you an answer. There were two populations. The average heights of those populations are slightly different.

When you get a bimodal frequency distribution chart, try to identify the two populations. Then determine whether you need to eliminate one of them. The less variability you have in your processes, the easier it is to stay within the specification.

SUMMARY

Frequency distribution charts are also called histograms. They are handy tools for evaluating processes.

The more common frequency distribution patterns include:

- Normal distributions that show the process is doing well (Figures 23.1 and 23.3)
- Normal distributions that show the process is in trouble (Figure 23.2)
- Normal distributions that show the process capability is much tighter than the specification (Figure 23.3)
- Truncated distributions that indicate the parts are being given 100 percent inspection before being shipped (Figure 23.4).
- Bimodal or two-peak distributions that show the product comes from two separate populations. Two-peak distributions should be investigated if the cause of the different peaks is not known (Figures 23.5 and 23.6).

24

Interpreting Control Charts

Process control charts are popular tools for monitoring and controlling quality. They reassure you when things are going well and they warn you when things are turning sour. With some companies, they also help production operators communicate with management.

In order to make them more mysterious, some quality organizations refer to these charts as **X**-bar and **R** charts. **X**-bar or $\overline{X}$ are statistical abbreviations for average; the charts use averages. **R** is the statistical abbreviation for range; the charts use ranges. The conventional symbols for average range are either **R**-bar or $\overline{R}$.

The charts we discuss are also called Shewhart charts after the man who invented them.

Other types of process control charts exist but they are not covered in this book.

Figure 24.1 shows a control chart with a "happy days" pattern; the process is in control.

Each point on the upper chart is the average of a number of measurements; in this instance, the number is five. Why five?

There is nothing magic about the number five. We could use three or four or even six. It is easier, however, to calculate the average of five measurements. Ten numbers might be even simpler but it takes longer to accumulate ten observations.

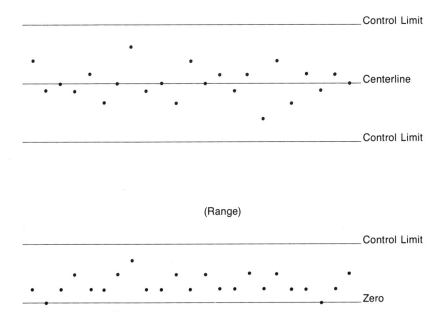

Control Limit

Centerline

Control Limit

(Range)

Control Limit

Zero

Figure 24.1 Control chart for a process that is in control (happy days pattern).

Specification limits should not be shown on the charts; they cause too much confusion. Specifications usually refer to individual parts, not group averages.

Control limits are shown on the chart to identify warning boundaries. The upper control limit is three standard deviations above the mean. The lower control limit is three standard deviations below the mean.

Does the range between the control limits sound like the process capability?

It should, but there are two differences.

1. The process capability is based on individual measurements; the control limits are based on averages. If there are five measurements within each group, the process capability range should be 2.24 times greater than the spread between the two control limits.
2. Process capability studies are normally run under ideal conditions; control limits are usually derived from plant data.

If the averages fall outside the control limits, the process is out of control; corrective action should be initiated. This should not happen for more than three groups out of a thousand without an assignable cause.

Procedures for calculating the control limits will be shown at the end of this chapter.

The **R** in **X**-bar and **R** charts refers to range. The range of the five measurements making up the average is shown in a second chart; it is normally plotted below the "average" chart.

Figure 24.1 shows a process where all of the averages and ranges on the charts are within the control limits. The process, therefore, is in control and management can rest easy. If the charts are maintained on the shop floor, management can see how things are going without talking to the operators; there is no need to disrupt production when everything is running smoothly.

Figure 24.2 shows an "out-of-control" or "yo-yo" pattern; the averages go up and down like a yo-yo. When this happens, production is normally making excessive defects. Scrap and rework follow.

With the yo-yo pattern, the range chart often gives you an out-of-control warning before production starts making scrap.

In Figure 24.2, the range chart is out of control. This, in turn, has led to the averages being out of control. The high variability, shown by the range chart, may have resulted from one of the following events:

- The control limits may have been established improperly.
- The raw materials may have become more variable.
- The process may have been changed.
- The process may have been restarted after a long period of inactivity.
- Experienced operators may have been replaced by trainees; this could be due to turnover or promotion.
- Manufacturing equipment or tooling may have worn out.

The chart shows two groups above the control limits and one below. The out-of-control condition, however, might have been predicted before the first average fell outside the control limits. The

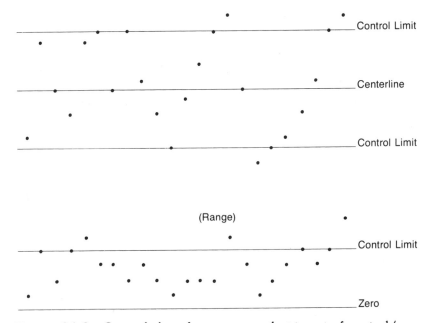

Figure 24.2   Control chart for a process that is out of control (yo-yo pattern).

range chart showed two groups on the control limit for ranges and one above it. In addition three groups were on the upper control limit for averages before one exceeded it.

Figure 24.3 has an "early warning" or "trend" pattern. It shows how control charts can be used to predict problems and initiate corrective action before the process goes out of control.

When trends are as pronounced as they are in Figure 24.3, special procedures are used to calculate the control limits. The limits are not based on average range data.

An early warning pattern demonstrates how process control charts can help identify trends. With this pattern, it is easy to initiate corrective action before the process starts making scrap.

Many manufacturing operations are vulnerable to tool wear. As the tooling gets dull, the dimensions of the product begin to drift; Figure 24.3 is a typical control chart for this type of operation.

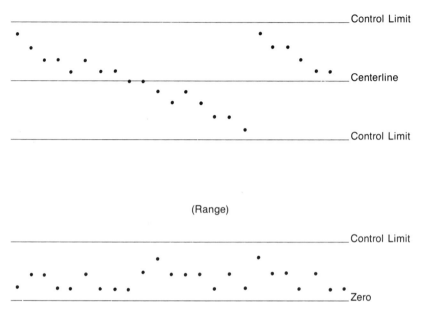

Figure 24.3 Control chart for a process that has tool wear (early warning or trend pattern).

By watching the chart, plant operators can anticipate when a tool will be dull enough to cause the parts to fall out of specification. At this time, the operators are able to shut down and change the tooling. Without control charts, they might make a number of discrepant parts before realizing they need a new setup.

When the product's dimensions tend to drift, as is shown in Figure 24.3, operators learn to make their setups to the high side of the control range. They continue running until the dimensions approach the bottom side of the control range. Then, after changing their tooling, they make a new setup toward the high end of the range.

Figure 24.3 illustrates a process where tooling wear causes the dimensions to drift downwards. Other processes, however, may drift upwards when tooling wears, producing an effect the opposite to the one shown in Figure 24.3.

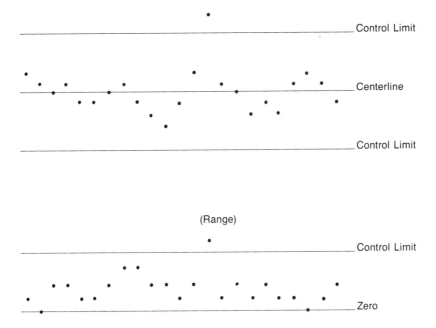

Figure 24.4 Control chart for a process that has unexpected problems (oddball pattern).

When process drift is the only problem, the range chart usually shows the process in control.

The "oddball" pattern in Figure 24.4 shows how control charts help identify unexpected problems.

In the beginning, the process represented by Figure 24.4 looked like it was in control. Then, all of a sudden, one group of measurements exceeded the upper control limit; defects were being generated.

Oddball groups should be investigated. They may be random, one-of-a-kind occurrences. On the other hand, they may be symptoms of larger problems.

The fact that the range exceeded the control limit at the same time as the average implies that the problem in Figure 24.4 was caused by one or two parts.

The reason for the oddball group may never be discovered by management unless the operator noted an unusual event on the control chart at the time it happened. Engineers and managers may guess for hours, but the operator usually knows what caused the problem. A short note on the chart is often more useful than hours of analysis and worry.

The "jump-shift" pattern shown in Figure 24.5 is frequently found when raw materials change.

Changes in the lots of oxidizer during the manufacture of solid rocket propellant often cause this type of pattern on ballistic control charts. In order to reduce the impact of these oxidizer changes, formulation engineers can:

- Blend 12 or more lots of oxidizer. Changes in 12-lot blends occur one-twelfth as often as changes in individual lots. In

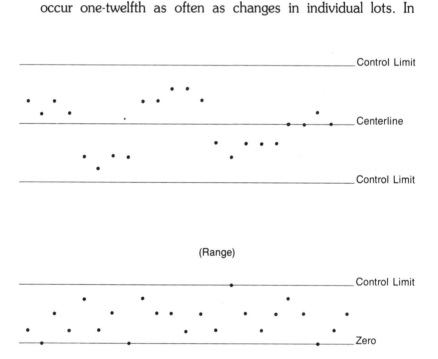

Figure 24.5 Control chart for a process with sudden changes (jump-shift pattern).

addition, successive 12-lot blends have about one-third the variability of successive lots.
- Pretest the material by using it in a batch of propellant before putting the new blend into production.

If you have a jump-shift pattern and do not know why, find out! It is difficult if not impossible to compensate for unidentified variables in a process.

The "why plot?" pattern, shown in Figure 24.6, identifies a process that is running so well that the control limits are never approached. If a process is doing that well, why spend time plotting the data?

"Why plot?" patterns usually occur when processes have been improved but the control limits have not been brought up to date. Control limits should be reviewed periodically and adjusted as necessary.

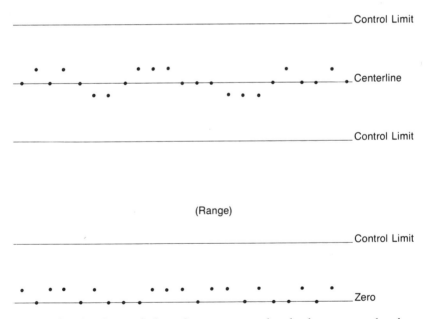

Figure 24.6 Control chart for a process that looks too good to be true (why plot? pattern).

Control charts were developed to identify and anticipate problems. If the probability of problems developing in a process is small, control charts may not be needed. The exception to this rule, however, comes when the variable being monitored is critical. Some variables, when discrepant, can result in customers being injured. These variables require monitoring; control charts are one technique for ensuring that the controls are in place and working.

HOW TO CALCULATE CONTROL LIMITS

The control charts shown in Figures 24.1 through 24.6 have:

- Upper control limits for the averages
- Lower control limits for the averages
- Upper control limits for the ranges

Their formulas are:

Upper control limit (averages)   = mean + .577 × average range
Lower control limit (averages)   = mean − .577 × average range
Upper control limit (range)   = average range × 2.115

In these formulas, "mean" refers to the average of all the measurements used in calculating the control limits.

The factors .577 and 2.115 only apply when the control charts are designed with five measurements per plotted point.

Example 24.1

Calculate the control limits for the resistance on an electrical resistor production line, given the following:

Average resistance   = 10,000 ohms
Average range (**R**-bar)   =   800 ohms
Measurements in each average =   5

Solution 24.1

Upper control limit (average)   = mean + (.577 × **R**-bar)
                                = 10,000 + (.577 × 800)
                                = 10,000 + 462
                                = 10,462 ohms

Lower control limit (average)   = mean − (.577 × **R**-bar
                                = 10,000 − 462
                                = 9,538 ohms

Upper control limit (range)   = 2.115 × **R**-bar
                                = 2.115 × 800
                                = 1,692 ohms

SUMMARY

Process control charts are excellent tools for discovering problems. They help identify:

- When a process is in control (Figure 24.1)
- When a process is out of control (Figure 24.2)
- When new process capability studies may be required (Figure 24.2)
- When specifications and process controls may require reevaluation (24.2)
- When a trend is about to cause problems (Figure 24.3)
- When unexpected problems occur (Figure 24.4)
- When there are changes in the process (Figure 24.5)
- When control charts may not be necessary (Figure 24.6)

25

Controlling Processes with Pre-Control

Some writers claim that production operators are the most important parts of a process. They stress bringing operators to their full potential. Given the right design, adequate equipment, appropriate tools, good materials, and proper training, good operators will make quality products. Quality is made or lost on the production line.

Pre-Control helps achieve the above objectives; it lets production personnel control the process without having to go through management, supervision, and engineering. It saves the time that is lost while the operators wait for an engineer or supervisor. During this time, time, the shipping bins may be filling with scrap.

The Pre-Control system was developed by Rath and Strong, Inc. It is easy to initiate and simple to use.

Advantages of the Pre-Control system include:

- It requires little or no paperwork.
- Once it is working, it requires no calculations.
- It lets operators control their own process.
- It is statistically sound but can be used by operators who hate statistics as much as you do.
- It helps operators make better setups.
- It helps identify changes in the setup and the process capability.
- It can be used with go/no-go gages.

Limitations of the system include:

- It assumes the measurements are distributed normally.
- It requires that the specification range is at least 10 percent broader than the process capability.

MECHANICS OF PRE-CONTROL

Steps required for Pre-Control include:

1. Dividing the specification into zones
2. Setting up the operation
3. Controlling the process

DIVIDING THE SPECIFICATION

Figure 25.1 shows the effect of dividing the specification into four equal areas.

The two middle areas in Figure 25.1 make up the target zone. This zone should be more than 3 SD wide; that is, 1½ SD on each side of the specification's midpoint.

If the process capability is the same as the specification and the setup is perfect, 86.6 percent of production will fall in this target zone. Pre-Control, however, requires that the specification range exceed the process capability by at least 10 percent. Therefore, more than 86.6 percent of production should be on target.

The two caution zones in Figure 25.1 fall between the target zone and the specification limits. The caution zones give you an early warning so that corrective action can be taken before you start making scrap. In a way, they are like the corrugated cement warning strips on some highways; they get your attention before you get in serious trouble.

SETTING UP THE OPERATION

With most operations, setup is an art. Production starts as soon as the operator or setup specialist feels the setup is right.

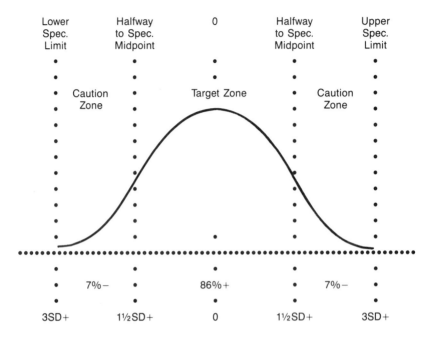

86.6% of all values fall within +/− 1½SD of the mean.
99.7% of all values fall within +/− 3SD of the mean.

Figure 25.1 How Pre-Control uses the normal curve.

With Pre-Control, the setup is less subjective. Five parts are pro-duced and measured. If any one of the five has measurements out-side the target zone, the setup procedure is repeated.

Once the measurements of five successive parts fall within the target zone, the operator starts running the equipment. This ensures that the setup is close to the center of the specification. If only a single part were checked, the setup could be marginal.

With Pre-Control, setup is not a serious problem when the specification range exceeds the process capability by more than 10 percent. It becomes costly and time consuming, however, when the specification range is smaller than the process capability. Under these conditions, it takes many setups before the operation qualifies for production. In addition, the "warning strips" or caution zones lose their effectiveness if they are sounding an alarm continously.

CONTROLLING THE PROCESS

Once operators produce five consecutive parts in the target zone and start running, they measure additional parts at predetermined intervals. A common interval is one hour. If inspection is time consuming, however, or the process has a history of changing slowly, longer intervals may be selected. When testing is relatively easy and the process has a history of instability, shorter intervals may be chosen.

The operators take two consecutive samples and check the first. Then they proceed according to the following rules:

1. If the first sample is in the target zone, the process continues to run until the next sample is due.
2. If the first sample is out of specification, the operator shuts down the process and tries to identify the problem. When the problem is too difficult for the operator, the supervisor is asked to help.
3. If the first sample is in the caution zone, the operator checks the second sample.
4. If the second sample is in the target zone, the process is allowed to run until the next sample is due.
5. If the second sample is in the caution zone, the operator makes a new setup.
6. If the second sample is out of specification, the operator shuts down and corrects the problem.
7. If a sample is out of specification, it is necessary to check the last eight pieces preceding the discrepant one. If all eight are in specification, the interim production can be accepted. If any of the eight is out of specification, all production between the discrepant unit and the previous sample should be inspected.

Problem 25.1

The operator is running a process that cuts cans to the proper height.

- The process capability is 0.09 inches.
  - a. Would the process qualify for precontrol?
  - b. What is the target zone?
  - c. What are the caution zones?

Answer 25.1

a. The specification range   = 4.06 − 3.94
                                              = .12 inches
    Process capability (given)   = .09 inches

Since the specification range is more than 10 percent greater than the process capability, the process should qualify for precontrol.

b. Specification range / 4   = .12 / 4
                                        = .03 inches
    Specification midpoint   = (4.06 + 3.94)/2
                                        = 4.00 inches
Target limits = midpoint ± (spec. range)/4
    Lower target limit   = 4.00 − .03
                                       = 3.97 inches.
    Upper target limit   = 4.00 + .03
                                       = 4.03

c. Caution zones are between the specification and the target zone. Therefore:

    Lower caution zone (bottom)   = 3.94 (spec.)
    Lower caution zone (top)   = 3.97 (target)
    Upper caution zone (bottom)   = 4.03 (target)
    Upper caution zone (top)   = 4.06 (spec.)
Figure 25.2 shows the answer in chart form.

Problem 25.2

After making the setup, the operator tested the first 5 cans that were produced. The results were: 4.00, 4.02, 3.99, 4.02, 4.05. What should the operator do now?

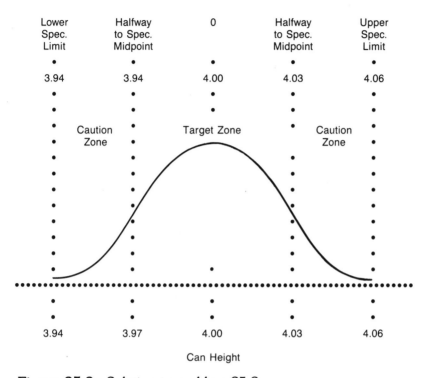

Figure 25.2 Solution to problem 25.2.

Answer 25.2

The operator should do another setup. The last point (4.05) is in the caution zone.

Problem 25.3

The operator did a second setup and processed an additional 5 cans. Their heights were: 4.00, 4.01, 4.00, 4.02, 3.99. What should the operator do now?

Answer 25.3

Start up the process and continue running until a problem is noted or another sample is required.

Table 25.1 Pre-Control in Action

Time	Measurement	Action by operator	Remarks
0800		Set up machinery	
0810	4.00, 4.01, 4.00 3.99, 3.96	Make second setup	Last height was in caution zone
0820	4.00, 3.99, 4.01 4.02, 4.00	Start operation	All heights in target zone
0900	4.00	Continue running	Height in target zone
1000	3.96	Check second sample	Height in caution zone
1005	3.99	Resume operation	Second height in target zone
1100	4.04	Check second sample	Height in caution zone
1105	4.05	Make another setup	Two heights in caution zone
1115	4.00, 3.99, 4.00 3.99, 4.01	Resume operation	All 5 heights in target zone
1200	4.05	Check second sample	Height in caution zone
1205	3.96	Make another setup	Second height in caution zone
1220	3.98, 4.00, 4.00 4.01, 4.02	Resume operation	All 5 heights in target zone
1300	4.10	Make another setup. Check previous 8 parts.	Height out of specification
1330	4.00, 3.99, 4.01 4.01, 4.01, 4.02 4.00, 3.99	Screening not necessary	All 8 parts in specification

Problem 25.4

The operator runs for an hour and then takes a sample. The height of the test can was 3.95 in. What should the operator do now?

Answer 25.4

Take another sample. 3.95 in. is in the caution zone. If the second sample is in the target zone (3.97–4.03), continue running. If the second sample is outside the target zone, shut down and do another setup.

Table 25.1 goes through the steps taken during a typical precontrol run. Review the steps, the measurements, and the operator's reaction to those measurements. The rules applying to the operator's actions should also be noted.

SUMMARY

Pre-Control is a system where operators control their own processes. Benefits that can be gained through implementing these procedures include:

- Improved morale.
- Reduced scrap.
- Better quality.
- Increased profits.

Pre-Control is worth trying.

26

Divide and Conquer

Some problem-solving exercises are like looking for a needle in a haystack; they are frustrating, time consuming, and the outcome depends on luck. Problem resolution, using the divide-and-conquer technique, is like a science; it is efficient, rewarding, and the outcome depends on the skill of the problem solvers.

The divide-and-conquer technique is one way of approaching problems systematically. Divide-and-conquer can be illustrated by applying it to a popular guessing game.

As an example, suppose you were asked to identify a whole number between 1 and 50,000. If you used the needle-in-a-haystack approach, you would average 25,000 guesses before choosing the right number. If you used the divide-and-conquer technique, you wouldn't need more than 16 guesses.

Say the number was 11,111.

Using the divide-and-conquer technique, your first question would be:

"Is it less than 25,000.1?"

The figure 25,000 is arrived at by dividing 50,000 by 2. The .1 is added to keep you from selecting a number on the dividing line. If the division does not result in a whole number, there is no need for adding the .1.

Subsequent guesses would be:

2. "Is it less than 12,500.1?" (25,000/2)
3. "Is it less than 6,250.1" (12,500/2)

4. "Is it less than 9,375.1?" (6,250 + 12,500)/2
5. "Is it less than 10,937.5?" (9,375 + 12,500)/2
6. "Is it less than 11,718.5?" (10,937 + 12,500)/2
7. "Is it less than 11,327.5?" (11,718 + 10,937)/2
8. "Is it less than 11,132.1?" (10,937 + 11,327)/2
9. "Is it less than 11,034.5?" (11,132 + 10,937)/2
10. "Is it less than 11,083.1?" (11,034 + 11,132)/2
11. "Is it less than 11,107.5?" (11,083 + 11,132)/2
12. "Is it less than 11,119.5?" (11,107 + 11,132)/2
13. "Is it less than 11,113.1?" (11,107 + 11,119)/2
14. "Is it less than 11,110.1?" (11,107 + 11,113)/2
15. "Is it less than 11,111.5?" (11,110 + 11,113)/2
16. "Is it 11,111?"

It has to be 11,111 because there is no other whole number that is less than 11,111.5 and more than 11,110.1.

The divide-and-conquer technique eliminates half of the available numbers with each guess. After the fifteenth guess, there is only one available number. The other 49,999 have been eliminated.

The divide-and-conquer principle also helps you solve production problems. Take the following example.

• The plant has production lines A and B.
• Both lines produce the same parts using the same four stage process.
• Line A produces 20 percent defectives.
• Line B produces no defectives.
• At each stage there are four variables capable of causing the defects.
• Which stage in the process is causing the problem?
• Which variable within that stage is responsible for the defects?

There are 16 possibilities (4 stages × 4 variables per stage).

You could run 16 tests and change a different variable each time. If you produced 20 parts during each test, you would make 320 pieces before you were through testing. If you used fewer than 20 parts per test, you might not generate enough data to solve the problem.

The needle-in-a-haystack approach is expensive. As an alternative, you could use the divide-and-conquer technique. This procedure would reduce the number of tests to 8 and the number of parts to 160. You would cut your variable costs by 50 percent and still get the same amount of data. How?

1. Run 20 parts through stages 1, 2, and 3 of line B. Then finish them at stage 4 of line A. If problems develop, stage 4 is at fault; it is the only section of the "bad" line that was used.
2. Run 20 parts through stages 1 and 2 of line B. Then switch them to line A for stage 3. Switch them back to line B for stage 4. If the number of defects is high, stage 3 is at fault; it is the only section of the "bad" line that was used.
3. Run 20 parts through stage 1 of line B and stage 2 of line A. Then switch them back to line B and finish steps 3 and 4. If defects increase significantly, stage 2 is at fault. Again, it is the only section of the "bad" line that was used.
4. Run the last 20 parts through stage 1, line A. Then switch them to line B and complete steps 2, 3, and 4. If discrepancies increase significantly, stage 1 is at fault; it is the only section of the "bad" line that was used.

Figure 26.1 gives you a graphical illustration of the divide-and-conquer strategy. If stage 2 is causing the problem, as Figure 26.1 implies, you would expect to find approximately four defectives during the third test. The four defectives represent 20 percent of the 20 test pieces run through the "bad" stage.

The above tests identified the step that caused the defects. The job is half done.

Now, make 4 runs of 20 parts each on line A. Change a different variable at the "problem" stage during each run. When you are through, the problem variable should be obvious.

Should be?

Yes. There is always the possibility that the defects won't show up during the test regardless of which testing system was used. This possibility is unlikely, however, unless the defects have been appearing and disappearing in an unpredictable manner. Problems that come and go with the phase of the moon are the hardest to resolve.

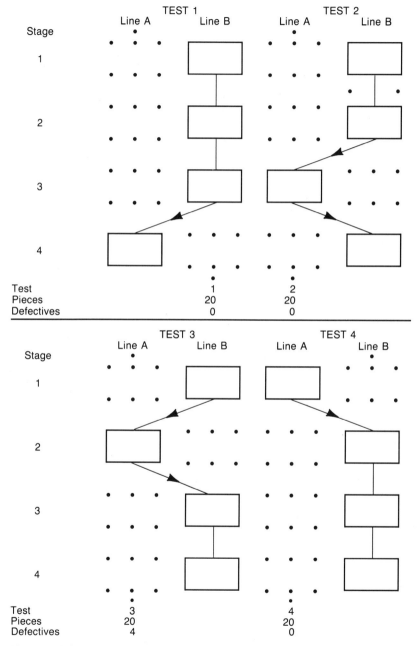

Figure 26.1 Divide-and-conquer test strategy.

SUMMARY

"Divide and rule" is an ancient political maxim cited by Machiavelli and used effectively by many strategists. When rephrased as "divide and conquer," it becomes an effective tool for solving production problems.

Why use the needle-in-a-haystack approach to solve problems when more efficient methods are available?

27

Understanding Regression Analyses

Prior to the age of computers, regression analyses were tools that seldom got out of the hands of quality engineers and statisticians. Few people understood what they meant; fewer could handle the tiresome math. Large quantities of meaningful information became lost in departmental files and company archives, never to see the light of day.

Regression analysis is a statistical tool used to determine the effect one variable has on another. Now that they are computerized, these analyses can be run by the average clerk.

This chapter helps you interpret the analyses and avoid the booby traps. It doesn't go into the calculations; most of the calculations will be handled by computers. This chapter doesn't teach computer programming, either; most of the programs will be written by professional programmers.

Traps to avoid when running regression analyses include:

1. Computer programs that give the wrong answers
2. Implied relationships that don't exist

CHECKING COMPUTER PROGRAMS

The easiest way to check a computer program is to give it a problem that has an obvious answer. Figure 27.1 is an example. It shows a perfect relationship between two sets of numbers.

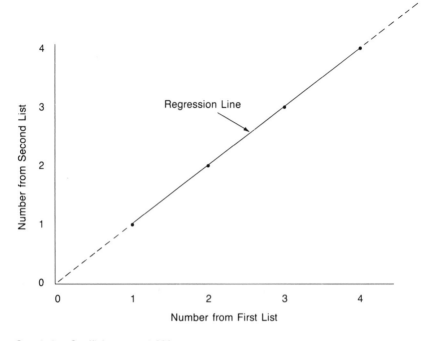

Correlation Coefficient   = 1.000
Probability that r is not 0 = 100%
Prediction formula:
    Number from Second List = 0 + (1 × Number from First List)
For those not allergic to statistics:
    Y = 0 + (1 × X)
Where X = Number from first list
        Y = Number from second list

Figure 27.1 Regression analysis (checkout pattern 1).

Data entered into the computer for "checkout pattern number 1" were:

First set of numbers	Second set of numbers
(X)	(Y)
1.0	1.0
2.0	2.0
3.0	3.0
4.0	4.0

Each number from one list is paired with an identical number from the other. As a consequence, if you pick a number from the first list you can predict the number in the second. Your prediction should never be wrong.

The graph in Figure 27.1 is called a scatter diagram. It plots each point using a number from the first list as the horizontal coordinate and its partner from the second list as the vertical coordinate. The first point has a horizontal value of 1.0 and a vertical value of 1.0. The second point has a horizontal coordinate of 2.0 and a vertical coordinate of 2.0. The numbers in the pairs always match.

Figure 27.1 is a "perfect" relationship because a straight line, called a regression line, can pass through every point. When it passes through all points and the slope is positive, its correlation coefficient r is 1.0. If the slope is negative, the correlation coefficient is −1.0.

The coefficients range between −1.0 and +1.0. Computer programs that give values outside these limits are wrong and should be discarded. The only exception to this rule involves roundoff errors. As an example, the computer may indicate that r = 1.000001. The answer is still wrong, but most people can tolerate a .000001 error.

The second thing to check is the slope. The line in Figure 27.1 rises one unit vertically with each unit it moves to the right. This means the slope of the line is +1.0.

The slope of the line will normally be identified in the computer printout as slope. It may also be called b, which is the abbreviation for slope used in prediction formulas.

The third thing to check is the intercept. The intercept is the value of the vertical coordinate of the regression line when the horizontal coordinate is zero. In most regression formulas, the intercept is called a.

In figure 27.1, the value of the intercept is 0. That means the regression line passes through the point where both the horizontal and vertical scales equal 0.

The fourth thing to look for is "the probability that the analysis is meaningful." Since we have ideal data, independent of reality, the probability in this case should be 100 percent.

It is difficult for statisticians to express this probability in English. They use terms like:

• The probability that r is not 0.
   (The probability that r is meaningful)

- The probability that r is 0
  (The probability that r is a dud)

The sum of these two statistics is always 100 percent. Both phrases are used by statisticians to say whether the analysis can be used with confidence.

The last thing to look for is the prediction formula. In Figure 27.1, the prediction formula is:

Number from second list = 0 + (1 × number from first list)

This formula follows the equation $Y = a + (b \times X)$ where:

- X = the number from the first list
- Y = the number from the second list
- a = the Y-intercept
- b = the slope of the regression line

Data shown in Figures 27.1 through 27.8 were run through a computerized regression analysis program. Results of the analyses are included beneath the scatter diagrams. Each figure illustrates a specific principle or problem.

Figure 27.2 presents a second set of data that can be used to verify a regression-analysis computer program.

Data entered into the computer for "checkout pattern number 2" were

First list of numbers	Second list of numbers
(X)	(Y)
1.0	4.0
2.0	3.0
3.0	2.0
4.0	1.0

Points were plotted like they were in Figure 27.1. The first point has a horizontal value of 1.0 and a vertical value of 4.0. The second point has a horizontal value of 2.0 and a vertical value of 3.0.

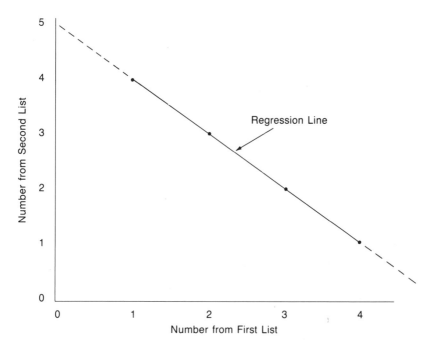

Correlation Coefficient   = −1.000
Probability that r is not 0 =   100%
Prediction formula:
    Number from Second List = 5 − (1 × Number from First List)
For those not allergic to statistics:
    Y = 5 − (1 × X)
Where X = Number from first list
        Y = Number from second list

Figure 27.2 Regression analysis (checkout pattern 2).

The regression line in Figure 27.2 drops one unit for every unit it moves to the right. Therefore, the slope is −1.0 or b = −1.0.

The scatter diagram in Figure 27.2 shows that this is a perfect relationship: A straight line can pass through all of the points. Therefore, the correlation coefficient r is −1.0. The sign of the correlation coefficient is always the same as the sign of the slope.

What is the intercept in Figure 27.2?

If you extrapolate the straight line to the left one unit, the vertical scale will show 5.0 when the horizontal scale shows 0.0. Therefore, the intercept is 5.0 or a = 5.0.

You don't have to understand the above explanation to check out a regression program. Just enter the data used in Figures 27.1 and 27.2. If you get the same answers, you can be relatively sure the computer program is all right.

If you want a more complete test, use larger sets of data containing both negative and positive numbers. The results of this analysis can be checked with other programs like those on the company's main computer.

Statistical programs should always be checked. People who rely on computer programs that haven't been tested are like purchasing agents who sign blank checks; they have more faith than common sense.

Regression analyses can help you; they can also hurt you. Figure 27.3, for example, is misleading.

Data entered into the computer for the "fool's paradise pattern" were:

First list of numbers	Second list of numbers
1	1
4	4

The important lesson to learn from Figure 27.3 is never to run a regression analysis on two points; you can always draw a straight line through them.

Since you can draw a straight line through any two points, all two-point analyses have correlation coefficients of either -1.0 or $+1.0$. With only two points, however, you can't determine the probability that the analysis is meaningful.

The probability of an analysis being meaningful is affected by the amount of data used. For a given correlation coefficient, the more data you have the more confidence you can place in the analysis.

When plant data are used for the analysis, you should have at least 30 points. When data from controlled experiments are used you can get by with fewer.

Properly run regression experiments minimize variations in anything not included in the analysis. The fewer random changes

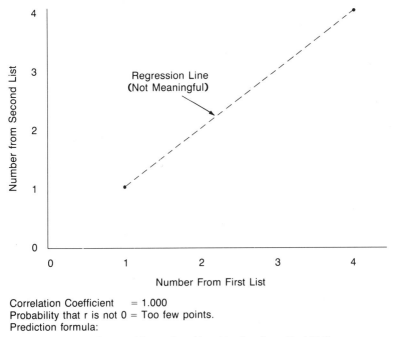

Correlation Coefficient   = 1.000
Probability that r is not 0 = Too few points.
Prediction formula:
    Number from Second List = 0 + (1 × Number from First List)
For those not allergic to Xs and Ys:
    Y = 0 + (1 × X)
Where X = Number from first list
        Y = Number from second list

Figure 27.3 Regression analysis (guaranteed perfection of fool's paradise pattern).

that take place, the better the analysis reflects the true relationship between the variables being studied.

MASSES OF DATA ARE NOT A CURE-ALL

Figure 27.4 shows that large quantities of data do not necessarily ensure a profitable analysis. There may be no interaction between the variables being studied.

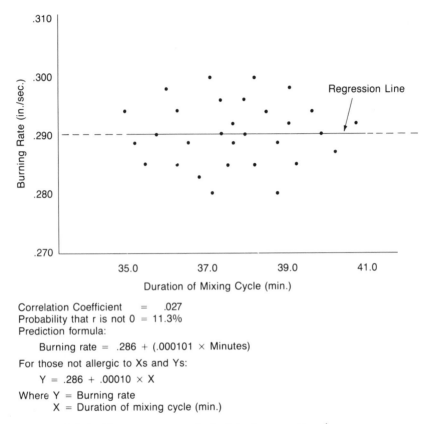

Correlation Coefficient   =   .027
Probability that r is not 0   = 11.3%
Prediction formula:

Burning rate = .286 + (.000101 × Minutes)

For those not allergic to Xs and Ys:

Y = .286 + .00010 × X

Where Y = Burning rate
      X = Duration of mixing cycle (min.)

Figure 27.4 Regression analysis (shotgun pattern).

Figure 27.4 has a scatter diagram with a "shotgun" pattern. The points look like they came from a shotgun when the target was almost out of range. Little or no interaction is shown between the propellant burning rate and duration of the mixing cycle. Analyses like this help identify what variables have to be controlled and which ones don't.

Propellant burning rates are important to the rocket industry. If the propellant burns too fast, it may create enough chamber

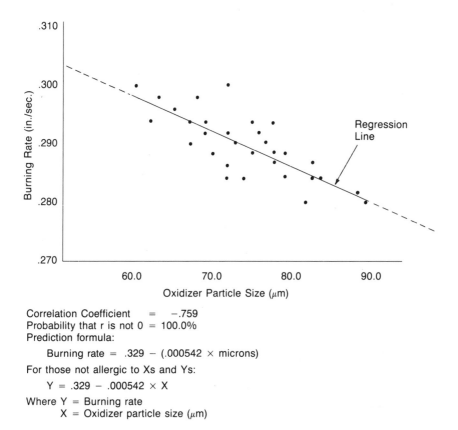

Correlation Coefficient   =   −.759
Probability that r is not 0 = 100.0%
Prediction formula:
     Burning rate = .329 − (.000542 × microns)
For those not allergic to Xs and Ys:
     Y = .329 − .000542 × X
Where Y = Burning rate
        X = Oxidizer particle size (μm)

Figure 27.5   Regression analysis (there's-safety-in-numbers pattern).

pressure to blow up the rocket. If it burns too slow, the rocket may not have adequate acceleration.

Figure 27.5 is a typical regression analysis between oxidizer particle size and burning rate.

Analysis of the data shown in Figure 27.5 yields a correlation coefficient of −.759; this is exceptionally good for an analysis of plant data. The probability that an interaction actually exists between the two variables rounds off at 100 percent (it is over 99.95 percent). The large number of points in the analysis and the excellent fit of the data contribute to this exceptional probability.

The formula for predicting burning rates using the same data is:

Burning rate = .329 – (.000542 × oxidizer particle size expressed in microns)

Based on the above formula, the formulation engineer would call for an oxidizer particle size of about 90 microns* if the target burning rate were .280 inches per second. The engineer would specify 60 micron oxidizer if the target burning rate were .296 inches per second.

OUTLIERS

Outliers or bad data can have a significant effect on regression analyses. This is illustrated by the "beware-of-outliers" pattern shown in Figure 27.6.

The data in Figure 27.6 are identical to data used in Figure 27.5 except for two points.

The 57 micron oxidizer was changed to 93 microns; the burning rate was not changed. This outlier represents a clerical error.

The 89 micron oxidizer was changed to 52 microns; the burning rate was not changed. This represents a second clerical error.

As a result of the two errors, the probability of the analysis representing a true interaction dropped from around 100 percent to 69.8 percent. As a consequence, users would have less confidence in subsequent predictions.

The changes in Figure 27.5 were made to illustrate two rules:

1. Never use a regression analysis without drawing a scatter diagram.
2. Beware of outliers or bad data.

One look at Figure 27.6 will tip off the trained statistician that the regression line does a poor job of predicting propellant burning rates; the slope of the line is biased by the two outliers. One outlier lowers the left end of the regression line; the other lifts the right.

---

* A "micron," or micrometer, is a unit of length equal to one millionth of a meter; it is abbreviated μm.

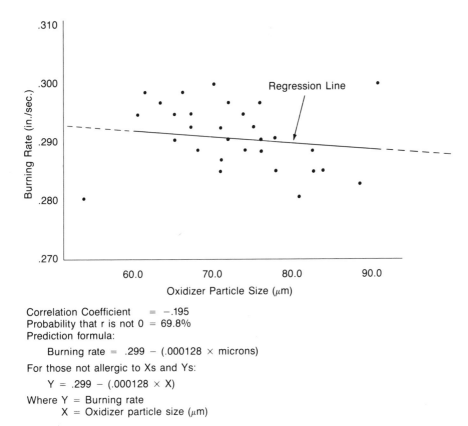

Figure 27.6 Regression analysis (beware-of-outliers pattern).

For 60 micron oxidizer, Figure 27.6 gives a predicted burning rate of .291 inches per second; Figure 27.5 gives a the prediction of .296, which looks more reasonable. For 90 micron oxidizer, Figure 27.6 gives a predicted burning rate of .287; Figure 27.5 gives a prediction of .280, which looks more reasonable also.

Figure 27.7 also has an outlier. In this case, however, the oddball point was intentionally entered to improve the predictions of the regression formula.

Thanks to a specially formulated propellant batch, data was generated well beyond the normal oxidizer particle size range

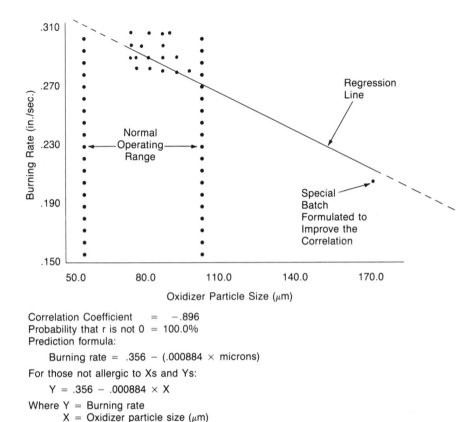

Figure 27.7 Regression Analysis (biased data pattern).

used for this specific propellant. This approach must be used with caution, if at all.

Regression analyses are influenced by points at the ends of the regression line more than they are by points in the middle. Thanks to the point to the right of the chart, the analysis shows a correlation coefficient of −.896. The calculated probability of there being a meaningful interaction between the variables approaches 100 percent.

The prediction formula may have been improved by adding data from the special batch of propellant. The correlation coefficient and confidence in the predictions, however, were inflated beyond reality.

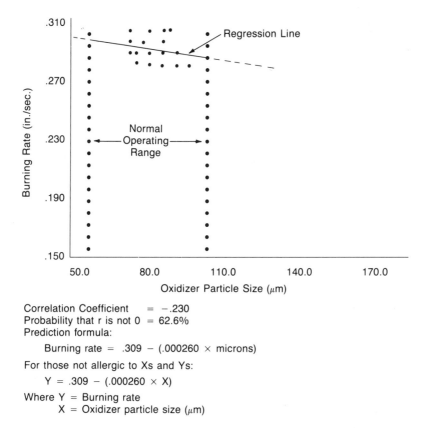

Figure 27.8   Regression analysis (unbiasing the data) (see Figure 26.7).

Figure 27.8 shows an analysis of the same data as Figure 27.7, except for the specially formulated batch of propellant; its data were removed. In this case, the correlation coefficient changed from −.896 to −.230. The probability of having a meaningful interaction between the variables dropped from almost 100 percent to 62.6 percent.

SUMMARY

Regression analyses are excellent for finding the relationships between variables. Precautions that have to be taken, however, include:

- Check out statistical computer programs with data that have been previously analyzed.
- Watch out for outliers or bad data.
- Never try to run a regression analysis on less than three points. With two, the correlation coefficient is always a perfect $+1.0$ or -1.0, even when no meaningful relationship exists.
- Eliminate as many variables as possible before running a regression analysis. As an example, if you are interested in how car speed influences gas mileage, don't use economy cars and gas guzzlers in the same analysis.
- Be wary of analyses that don't make sense. If you make enough analyses, you will eventually find one that appears significant, even though it isn't. An example might be relating the scores of probowl football games to movements of the stock market. You may get a good correlation; but don't let a pigskin determine your investment timing.

28

Designing Experiments

Claude Bernard (1813–1878), the founder of experimental medicine, wrote, "Observation is a passive science, experimentation an active science." With experimentation, you have to observe and plan. If you want to get your money's worth, you should also design your experiments.

At one time, most experiments involved making changes in one variable and observing their effect on another. As an example, you might adjust reactor temperature in an alkylation unit and observe its effect on alkylate quality.

Simple, two-variable experiments are helpful for verifying regression analyses of plant data. When you need information on additional variables, however, you should use more complex designs.

Expert statisticians have experimental designs that will confuse almost any manager. In this book, however, we will stick to simple methods. If they interest you, in-house experts can go into greater depth with flashing computers and charts galore.

The most important considerations in experimental design are:

1. Replication
2. Randomization
3. Interactions

REPLICATION

Replication is the statistician's word for duplication. Statisticians seldom use simple words when something more complex will do.

In simple English, replication means getting more than one measurement for each test condition.

Why go to that expense?

Replication helps you determine whether you left critical variables out of the experiment. If you don't get consistent results, you haven't controlled an important variable.

Figure 28.1 shows a simple experiment with replication.

The experiment in Figure 28.1 was designed to determine whether propellant burning rates were affected by:

- The method of mixing ingredients (premix)
- The particle size of the curing agent (psd)
- An interaction between the two

This experiment is relatively simple. Only three variables are being evaluated. Everything else is kept constant.

The experiment runs duplicate tests for each condition. In Figures 28.1 through 28.3, results for each test condition are isolated in their own private box. We will call these boxes test cells. As an example, the box in Figure 28.1 that shows .30 and .31 is the cell for tests using premix method 1 and 90 μm curing agent. The box that shows .36 and .35 is the cell for tests using premix method 2 and 90 μm curing agent.

The two numbers in the cell give the results of duplicate tests. In this case, they represent propellant burning rates expressed in inches per second.

Ranges are calculated for each cell to determine the variability of the process when all test conditions are held constant.

In this instance, all four ranges are .01 inches per second. As a consequence, the average of the ranges in the test cells is also .01 inches per second. The small range of data within the cells indicates that no uncontrolled variables interfered with the test; either they were part of the experiment or were kept constant.

		Premix Method		
		1	2	Mean
Curing	90	.30 .31	.36 .35	.33
Agent				
PSD	60	.32 .31	.37 .36	.34
(μm)	Mean	.31	.36	.335

Diagonal Average				Diagonal Average
.335				.335

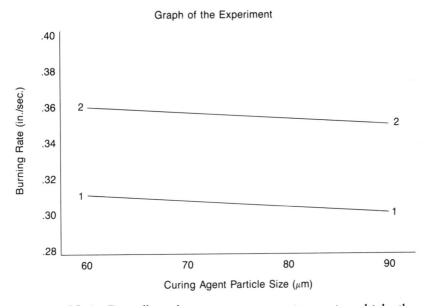

Graph of the Experiment

Figure 28.1 Propellant burning rate experiment in which the premix method dominates and there is no interaction.

Other conclusions that can be drawn from the test include the following:

- The burning rates changed very little when the particle size of the curing agent went from 60 to 90 microns. The difference in the averages was .01 inches per second (.34 − .33). This is the same as the average range within the test cells. Based on this data, it is unnecessary to worry about the curing agent particle size as far as burning rates are concerned.

- The average difference in burning rates for the two premix processes was .05 inches per second (.36 − .31). The premix processing procedures appeared to have a definite effect on the burning rates. This information would be critical to the formulations engineers; they would have to compensate by changing other variables whenever the premix procedure was modified.

The graph in Figure 28.1 gives a "picture" of the experiment. The two lines are almost level. This confirms the observation that the curing agent particle size had little effect on burning rates.

There was a fairly large distance between the parallel lines on the chart. This indicates that the premix process affected burning rates.

The lines don't converge, diverge, or cross. This indicates there is no significant interaction between premix methods and curing agent particle size.

RANDOMIZATION

Randomization means that tests in the experiment must be run in a random sequence. As an example, you might use the following order for testing:

Sequence	Test
1	90 micron curing agent and premix method 2
2	60 micron curing agent and premix method 1
3	60 micron curing agent and premix method 2
4	90 micron curing agent and premix method 1
5	60 micron curing agent and premix method 1
6	90 micron curing agent and premix method 2
7	60 micron curing agent and premix method 2
8	90 micron curing agent and premix method 1

The objective of randomization is to avoid having the tests ruined by time or sequence-related variables. For example, time of day could affect ambient temperature, humidity, alertness of the test team, or other unknown variables. These variables, in turn, might influence propellant burning rates.

Without randomization, you don't know for sure whether variables you didn't control had an impact on the experiment.

Figure 28.2 uses replication and randomization to show that critical variables were left out of an experiment.

In Figure 28.2, the average of the ranges within the cells is .05; this appears excessive. The calculation is:

Average range = (.08 + .04 + .02 + .06)/4 = .05

The difference between the averages for the two premixes is only .02; (.375 − .355). With such a small difference in premix averages and such a large average range, you can't tell whether the difference in burning rates for the two premix methods is due to chance.

The .28 burning rate with one of the "premix 1 and .50 grams per cubic centimeter (g/cc) catalyst" tests might have been due to it containing an oxidizer that was different from those used in the other tests. If the oxidizer had been kept constant, the value might have been .40 inches per second. This one modification would reverse the premix results; premix 1 would have shown the higher average burning rate.

To be sure, you would have to run another experiment and do a better job of controlling variables that were not in the test.

The difference between the averages for the two lots of burning rate catalysts was .07 inches per second (.400 − .330). This is larger than the average range of the data within the cells and indicates that the bulk density of the catalyst does affect the burning rates.

INTERACTIONS

Figure 28.3 shows the results of a microfilm vs. paper records experiment. It provides detail on interactions.

In this mythical experiment, we are evaluating whether either time of day (A.M. and P.M.) or type of records influence clerical

		Premix Method		
		1	2	Mean
Burning	.50	.28 .36	.32 .36	.330
Rate				
Catalyst	.30	.38 .40	.38 .44	.400
(g/cc)	Mean	.355	.375	.365

Diagonal Average				Diagonal Average
.365				.365

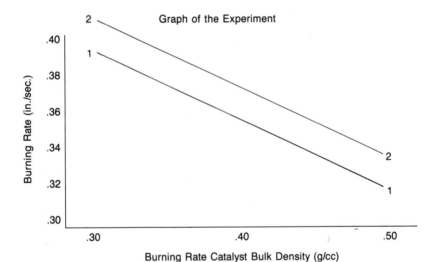

Figure 28.2 Propellant burning rate experiment in which the burning rate accelerator dominates and there is some interaction.

error rates. This test was initiated because clerks were complaining about having to work with microfilm.

In simple experiments like the ones we have been discussing, interactions can be identified two ways.

- By viewing graphs like those in Figure 28.3
- By analyzing "diagonal averages" shown on the charts

If the graph shows parallel lines, an interaction is improbable. If the lines converge, diverge, or cross, an interaction is probable. In this case, the lines cross, indicating a probable interaction.

The "diagonal average" at the right of Figure 28.3 is the average of data from the microfilm-**PM** cell and the paper-**AM** cell. The two cells are diagonally across from each other. This diagonal average equals 21.

The calculation is (22 + 26 + 18 + 18)/4 = 21.

The diagonal average at the left of Figure 28.3 is the average of data from the paper-PM cell and the microfilm-AM cell. Again, two cells in the average are diagonally across from each other. This diagonal average equals 18.

The calculation is (20 + 18 + 18 + 16)/4 = 18.

In this case, the difference between the two diagonal averages is three, (21 − 18 = 3).

The average of the ranges within the four cells is two.

The calculation is (4 + 2 + 2 + 0)/4 = 2.

Since the difference between the diagonal averages is larger than the average range within the cells, there is a probable interaction.

In statistics, however, you are seldom 100 percent sure of anything. If there were an interaction, it could be attributed to the clerks having problems with microfilm in the afternoon when they were tired. They may not have had problems with the microfilm in the morning when they were fresh.

Techniques for calculating the exact probability of an interaction are beyond the scope of this book. Computer programs for analysis of variance (ANOVA) or analysis of means (ANOM) can do the calculations for you if the need arises.

SUMMARY

Designed experiments help determine the effect variables have on a process. They help identify what you don't have to control as well as what you do.

In order to get the most out of an experiment, you should replicate, randomize, and look for interactions between variables.

Type of Records			
	Microfilm	Paper	Mean
Time	22 26	20 18	21.5
of Day AM	18 16	18 18	17.5
Mean	20.5	18.5	19.5
Diagonal Average			Diagonal Average
18			21

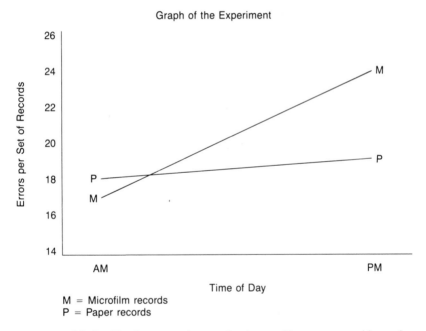

Graph of the Experiment

M = Microfilm records
P = Paper records

Figure 28.3 Evaluation of records, /microfilm vs. paper/ based on clerical error rates.

Replication involves running at least two tests for each set of conditions. If critical variables are left out of the experiment, replication will give a warning—duplicate tests will show inconsistent results.

Randomization helps keep cycles and time-related variables from causing misleading results. Be particularly careful that duplicate tests are not run one after the other.

Interactions are among the more difficult problems for novice trouble-shooters. Designed experiments, like those in this chapter, are effective in identifying interactions.

Techniques covered in this chapter will enable you to identify probable relationships. In most cases, you will not need to know the actual probabilities. If you do, however, they can be calculated by computer using either ANOVA or ANOM.

29

Plan Experiments—Don't Muddle Through

In **The Republic**, Plato wrote, "The beginning is the most important part of the work." This is especially true for designed experiments. In the beginning comes the plan. Without it, many experiments fail.

When planning an experiment, you should:

1. Set objectives.
2. Select the variables.
3. Determine the range for the variables.
4. Anticipate interactions.
5. Keep it simple.
6. Prepare for the worst.
7. Choose the right gages.
8. Calibrate gages.
9. Use trained technicians.
10. Check supplies.
11. Record details.

SETTING OBJECTIVES

Without an objective there's no need for an experiment.

At least three people should agree on the objective. This includes identifying the problem and summarizing expectations.

Those selecting the objectives should come from different departments and disciplines. Independent viewpoints often provide a catalytic effect, creating a better understanding of underlying problems.

The team should also consider whether the expected benefits are cost savings, scrap reduction, increased sales, fewer returns, more efficient operations, or reduced maintenance.

SELECTING VARIABLES

Make sure all possible causes of variation are considered. The probable causes should be part of the experimental design; other causes should be held constant, just in case they have a greater influence than was anticipated.

Include quality engineers, specialists, and production personnel in the selection process.

Quality engineers should have skills in experimental design. They should also have information on which variables to test. Much of this knowledge would come from regression analyses run on plant data. Additional ideas might come from discrepancy reports supplied by inspection personnel.

Specialists can identify candidate variables from a technical standpoint. If the problem involves metallurgy, include a metallurgist. If it involves an electrical problem, include an electrical engineer.

Operators and shop supervisors with extensive experience in plant operations should be included when production problems are involved. In many instances, closeness to the problem gives operators a special insight. At times, hands-on experience is more valuable than theory.

Operator opinions, however, can't be accepted as the absolute truth. This is particularly true with complex problems.

Production people occasionally run their own tests and develop firm opinions. Sometimes, however, they oversimplify and fail to detect interactions between variables. They can also fail to observe changes that happened during the test—changes that are more important than the adjustments being made. In addition, they seldom record or control all of the critical variables.

DETERMINE THE RANGE FOR THE VARIABLES

The range of the variable shouldn't be too big or too small. If too big, scrap will be excessive. If too small, the test may not be sensitive enough to identify meaningful relationships.

In some experiments, the variables have a range that matches the process capability for the product. The test, however, will be more sensitive when the range is broad enough to permit production of some discrepancies. Variables such as temperature, pressure, speed, and volume should be given as much leeway as possible in order to identify their limits in the process.

Short trial runs can be used to help estimate allowable ranges. This is particularly true when there are many variables. Trial runs also help identify probable interactions between variables.

During a test, the risks you take should be influenced by whether the end product will be sold. Other factors include whether it is easy to identify, remove, rework, or scrap discrepant production. Shipping schedules and lead times must also be considered. These concerns can be discussed with the production, marketing, sales, and financial departments before the test is run.

ANTICIPATE INTERACTIONS

Interactions are frequently missed by inexperienced personnel; they have to be considered. A typical case of interaction was briefly discussed in the previous chapter.

A hypothetical company wanted to evaluate the effect that microfilm had on the error rates of their clerks. In the morning, one set of clerks processed a difficult batch of microfilm records; another set of clerks processed a similar batch of paper records.

The test was repeated that afternoon. This time, the clerks who were on microfilm were switched to the paper records; the clerks who were on paper records were switched to microfilm.

The results were:

Total errors (microfilm records) = 32
Total errors (paper records)   = 30

It appeared there was little difference between the two types of records until the data were broken down by time of day. Then the data showed:

Time of day	Microfilm records	Paper records
Morning	12 errors	14 errors
Afternoon	20 errors	16 errors

Results of this test are shown in Figure 29.1.

Based on these data it might be concluded that there was an interaction between time of day and type of records. The clerks did well on microfilm in the morning when they were fresh. They made more errors with microfilm in the afternoon when they were tired.

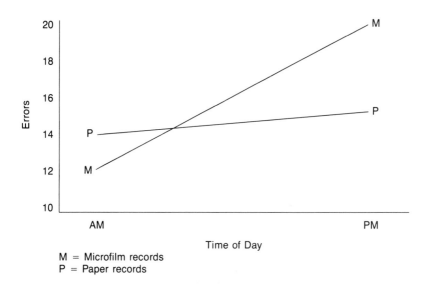

M = Microfilm records
P = Paper records

Figure 29.1 Designed experiment of the error rates for microfilm versus paper records.

KEEPING IT SIMPLE

When running experiments, simplicity is beautiful. The more complex the experiment, the more it will cost and the more it will confuse. Managers seldom accept anything they don't understand.

One of the formulation engineers at a West Coast solid rocket plant used complex procedures for controlling burning rates. No one in management could understand what he was doing; they were confused by his graphs and charts. The only reason they didn't transfer him was that he had the lowest scrap rate of any formulations engineer in the company. They understood the results; they didn't understand how he got them.

PREPARE FOR THE WORST

When getting ready for an experiment, estimate the number of observations that will be needed. Then be prepared to run more.

Experimental designs normally call for an exact number of test results. In the process of running the test, however, a variety of unexpected things may occur. Test items may be broken, damaged, lost, or just disappear. If you are prepared to run additional items, you can usually avoid having to rerun the experiment.

CHOOSE THE RIGHT GAGES

Choose the right measuring equipment. It must have enough accuracy and precision. How precise?

A rule of thumb is, "The device should be ten times as precise as the reported results." Then for measurements precise to .001 inch, the device should be able to measure to .0001 inch.

In practice the gaging accuracy should depend on the nature of the measurement and the consequences of inaccuracy. Determine these requirements during the planning phase of the test.

If new measuring devices are required, don't accept the claims of the manufacturer without checking the devices in your own laboratory. Manufacturer's claims are often written by nontechnical people who don't understand the terms involved or your use of the equipment.

It is also helpful to have your own technicians run tests on new equipment; it may require skills and training your people don't have.

When the desired precision can't be obtained from an available gage, run a larger number of measurements. Then average them. If you arrange your tests in groups of four, their averages will scatter half as much as the individual data. If you run groups of nine, their averages will scatter a third as much as the individual data. The improvements vary with the square root of the group size.

CALIBRATE GAGES

Before starting the test, calibrate your measuring equipment.

An operator stopped one test because the pressure readings were 5 psi too low. Another gage was brought in to replace it and everything came out right.

The first gage had been in service for a long time but the second came straight from the calibration lab. This minor crisis could have been avoided if the first gage had been checked before the test started.

Many companies have calibration stickers on their gages, giving the engineers an indication of how recently they were checked. Some calibration laboratories also keep extensive records on the gages they service. If many of the gages that come in need to be adjusted, the period between calibrations is shortened. If all of the gages that come in are functional, the period between calibrations is extended.

USE TRAINED TECHNICIANS

When planning an experiment, select trained operators and inspectors ahead of time. Trained backups should also be available. Scheduled personnel may get sick or pulled away from the job and the choice of substitutes should not be left to chance.

Periodically check inspection personnel used for designed experiments. There is always the possibility of unintentional bias or rusty skills influencing the measurements.

Problems increase when new equipment and gages are used. Special training for inspectors and their backups may be required before the test is started.

Unusual materials may also cause problems. This is especially true with semirigid materials. It is too easy to get distortion when "feel" enters into the measurement.

CHECK SUPPLIES

Controlled experiments usually require special supplies and equipment. Make sure they are available before the day of the experiment. It is wasteful to assemble the personnel and equipment for a test only to find that a drum of special material isn't available.

For large experiments, have a specific person responsible for the arrangements. This assignment can be simplified by having a check list so that nothing is left to chance.

Pretest meetings are also helpful. They enable production to give its final approval and test personnel to clarify assignments. At this time, some organizations conduct a dry run, talking through the operation step by step. The more complex the test, the greater the chance that a dry run will uncover potential problems that were not anticipated.

RECORD DETAILS

In the process of running the experiment, unexpected events often occur. Record them. At the time of the test, they may seem unimportant. After the data are analyzed, these unusual events may help explain inconsistent data.

Additional things to record include:

- Names of everyone involved with the test, including their assignments
- Line, equipment, and machine numbers
- Times the test started and finished.
- Time of any unusual events

- Temperatures and pressures
- Raw material lot numbers
- Feed rates

After the experiment is run and the data tabulated, determine whether unusual events influenced the test. This information should be included in the final report. Questions always arise after the test report has been issued. If details are hidden in the personal notes of individual test team members, they may get lost before their impact on the test is known. They may also get thrown out during the company's annual house cleaning. Information of this type is seldom needed until the day after it is thrown away.

SUMMARY

Designed experiments are an excellent tool for developing information on the best way to run a process. When accompanied by planning, they can be reliable and cost effective. When conducted without adequate preparation, however, they can be costly and subject to justifiable skepticism.

30

Traps, Snares, and Pitfalls Along the Way

The thing about quality problems is they will never be in short supply.

In days of yore, quality weapons could save your life. Today, quality products can save your company.

If you want to survive in business, you have to avoid the traps, snares, and pitfalls that can put you in a quality bind. Common hazards include:

1. The scapegoat trap
2. The let-the-quality-manager-do-it trap
3. The good-old-buddy trap
4. The fox-guarding-the-hen-house trap
5. The declaration-of-independence trap
6. The head-in-the-sand trap
7. The old-dog-new tricks trap
8. The time trap
9. The one-happy-family trap
10. The penny-pincher trap

The scapegoat trap is the most popular. When something goes wrong, find someone to blame. If this trap is set by an expert, everyone but the company can come out ahead, even the scapegoat.

One West Coast manager devised, directed, and implemented an experiment that failed. He changed a process without adequate planning and created scrap costing over $20,000. Rather than admit his error, he blamed a docile subordinate. In exchange for taking the heat, the subordinate had a job for life, or as long as the negligent boss survived.

Production is a common scapegoat. It is easy to blame production for problems due to poor design, defective raw materials, inattentive inspectors, inadequate training, or improper packaging.

As long as offenders have scapegoats, quality won't improve.

The let-the-quality-manager-do-it trap is just as bad. You can't delegate all quality functions to the quality manager and then forget them. Top management must accept responsibility for having a sound and effective quality program.

If scrap, rework, customer complaints, returns, recalls, and liability suits soar, top management is doing something wrong. It may have selected the wrong quality manager, slashed the wrong budgets, or created the wrong atmosphere. People are quick to assume that quality isn't important when management ignores it.

Retaining an incompetent quality manager is like letting gangrene spread because you don't want to loose a leg.

Where do incompetent quality managers come from?

The good-old-buddy trap is a possible answer.

Is the quality manager a "good old buddy?" How was the selection made? Improper responses include:

- "Joe's too old to work hard; give him an easy job."
- "Ginny did a good job in engineering; promote her."
- "Roberto is the boss' pal; he can cut red tape."
- "Harry's tops in the shop; he knows production."
- "Rose gets along with everyone; choose her."

These people might be outstanding managers if they have the necessary skills and training. Without those skills, they can spawn quality disasters like a dying salmon delivering roe.

Quality managers who know nothing about statistical quality control, vendor quality, specifications, auditing, and inspection can do more damage than a tidal wave hitting Hawaii during a surfers' convention.

The fox-guarding-the-hen-house trap usually involves having inspection report to production. The assumption is that this organizational structure will expedite corrective action.

The problems with the "fox" approach include:

- If schedules are tight, junk may get shipped
- Few production people understand problem analysis.

Although these shortcomings can be overcome, they seldom are. Exceptions occur when the production team is as knowledgeable and quality oriented as its quality counterpart.

Do production executives ever admit they are not quality oriented? Seldom.

Does the average production manager have skills in the technical phases of quality assurance? No!

Fortunately, however, there are exceptions. A few production managers can be trusted with the keys to the hen house.

The declaration-of-independence trap involves giving inspection independence—separating inspection from the quality department. Most inspectors are excellent at identifying defects and alerting supervision to the need for corrective action. Few inspectors, however, are skilled at summarizing, analyzing, and interpreting data. Fewer are proficient in statistical quality control.

The head-in-the-sand trap is based on the false assumption that the quality function is not important; the quality manager can report to any intermediate level executive, including the head janitor.

The flaw in this approach is that each level of management filters information. If the quality manager reports to the engineering vice president, most quality reports will have an engineering bias. If the quality manager reports to the manufacturing vice president, most quality reports will have a production bias.

One engineering vice president refused to acknowledge any of the accomplishments of trouble-shooters in a quality department that reported to him. He insisted that all problems had to be solved by design engineers.

Although his quality personnel documented their trouble-shooting activities in detail, the reports were never taken seriously. His loyalty to the engineering organization hampered his ability to use the quality organization effectively.

The old-dog-new-tricks trap is built on the assumption that production is an art. Some old timers claim, "No one without years of manufacturing experience can show production how to do a better job." They can't conceive that principles behind statistical quality control are universal. These principles can be applied to administrative, engineering, production, and design problems with remarkable effectiveness. "Cause-and-effect" factors in industry operate in a statistical manner. The inherent variation can be identified, evaluated, and controlled.

The time trap is built on the false assumption that being first in the marketplace is all that counts. When this philosophy is followed to the extreme, faulty design, customer rejection, product recall, and liability suits follow. If being first was all that counted, IBM's personal computer would have been a failure. In reality, the opposite was true. IBM was slow to develop a personal computer; when they did market a product, it soon became the industry standard. Most of the early leaders are now selling IBM-compatible models. IBM, however, has not had to copy their products.

The one-happy-family trap is built on the false assumption that major departments will work together without direction from above. This theory ignores the existence of departmental jealousies, empire building, and incompatible department heads.

How many parents would let five-year-old children determine the family diet? Most children would choose candy, cookies, and soda pop. Managers are more mature; they would probably eat right. Given freedom in setting priorities, however, they might revert to their youth. How many managers give top priority to helping solve another manager's quality problems, without a little executive guidance?

The penny-pincher trap is based on the false assumption that all overhead operations are expendable. When sales are slow and funds are limited, overhead operations are usually cut. Quality assurance is overhead.

Avoiding the traps, snares, and pitfalls of quality is like becoming a bomb disposal expert. You have to learn fast to grow old.

Customers are becoming smarter. Products are more complex. Competition is tougher.

Japanese, Koreans, and Yogoslavians are tough competitors; they learned to market quality products at competitive prices.

In order to keep up with the competition, the United States must market products the customers want with the quality they demand at a price they can afford. If we do otherwise, we will come out second best. Or third. Or fourth. The choice is ours.

31

Quotable Quotes

The time has come, the walrus said,
To talk of many things:
Of shoes—and ships—and sealing wax . . .
Of quality—and kings.

Our apologies to Lewis Carroll for rephrasing his lines from "Jabberwocky," but who wants to talk about cabbages?

This book has stressed the common sense side of quality assurance. Most of this work is based on the accumulated experience of two professional quality practitioners. We both spent decades trying to convert the art of quality control to a science.

Basic concepts contained in the book are summarized by the following quotations:

"Technical material in this book was written for executives. If they understand it, anyone can. If they read it, everyone will." (Preface)

"The good news is that we know how to economically manufacture high-quality, reliable products. The bad news is that many manufacturing companies have not learned the good news." (Chapter 1)

"The quality department's function is to help other departments coordinate their efforts to produce a quality product." (Chapter 1)

"A quality department's findings can be helpful if top management uses them. Otherwise, they have no more value than last week's cup of coffee, dehydrated by cigarette butts." (Chapter 1)

"Workers quit listening to slogans the moment they differ from management's actions." (Chapter 1)

"Never underestimate your customer's ability to think up cruel and unusual punishment for your product." (Chapter 2)

"Scientific sampling helps determine whether material conforms to the quality requirements; it does not control quality." (Chapter2)

"Blaming your quality department for poor quality is like blaming your secretary because you forgot your spouse's birthday." (Chapter 2)

"An executive's exposure to the technical aspects of quality control is similar to a mouse's exposure to a hungry python. If either one fails to learn the strengths, weaknesses, and language of the enemy, they become victims of their own ignorance. (Chapter 2)

"When making general purpose ash trays for skid row bistros, you have lots of leeway with the precision and accuracy of your measurements. When shooting the cigarette from the mouth of an attractive model, you have very little leeway." (Chapter 3)

"Without coordination . . . a problem-solving exercise resembles a Keystone Cop comedy, with everyone rushing off in different directions." (Chapter 3)

"When they (executives) can contribute, they don't get in the way as much." (Chapter 3)

"Some are more comfortable hiding from new ideas than going through the trauma of showing their ignorance to subordinates." (Chapter 3)

"Customers object to being inconvenienced by someone else's sloppy work." (Chapter 4)

"Now customers don't consider a product reliable unless it is usable when it goes out of style." (Chapter 4)

"Don't test the loyalty of your customers. Do preproduction testing instead." (Chapter 4)

"Customers seldom get compensation for the time, effort, postage, gasoline, shoe leather, and heartburn medicine consumed in their quest for replacements, even when covered by a warranty." (Chapter 5)

"Companies that don't ensure that their replacements are functional invite competition into their territory." (Chapter 5)

"Mix an injured customer, an opportunistic lawyer, and a consumer-oriented jury and you can have an adverse liability settlement that could drive you out of business." (Chapter 6)

"Virtually everything can become dangerous if it is defective enough." (Chapter 6)

"Customers . . . are as hard to predict, anticipate, and understand as hyperactive three-year-old children on a diet of chocolate bars and sugar snacks." (Chapter 7)

"Few people go out of their way to provide customer feedback unless there is 'something in it for them.'" (Chapter 7)

"Feedback from service and repair agencies has great value when the information is distributed throughout the company. It has little value when scanned and stored in someone's private file, even if that someone is the company president." (Chapter 7)

"Look for critical defects like a small town cop running a speed trap. Don't relax, no matter how long it's been since you found a potential catastrophe." (Chapter 8)

"It is embarrassing to reject a good incoming lot of material because your gages were out of calibration. It is even worse to accept a discrepant lot of your own product." (Chapter 8)

"Quality catastrophes are like land mines. If you anticipate their presence and plan to avoid them, they seldom blow up. After one explodes, it's easy to see what you should have done—if you live through the blast." (Chapter 9)

"If your product isn't designed correctly, there's nothing production can do to make it better." (Chapter 9)

"Distraction. Inspectors might be distracted by a miniskirt swishing by or by a hunk from Muscle Beach passing through on a tour." (Chapter 9)

"Many corporate takeover defenses hurt the company's employees by draining off funds that could be used to buy tools and equipment they need on the job." (Chapter 10)

"Sleep is fundamental. You can't get ahead if you aren't awake when promotions are being handed out." (Chapter 11)

"The quest for quality requires that each worker be treated as a unique individual." (Chapter 11)

"The bosses are not always wrong. Supervision and management is so complex that textbook solutions may not apply in many cases." (Chapter 11)

"Textbooks are full of reasons why you should praise workers when they do a good job. Few books consider the impact of praising them when they do a poor job." (Chapter 11)

"Automatic raises encourage mediocrity, not excellence." (Chapter 11)

"The sooner you find and correct these problems the less damage they do." (Chapter 12)

"The best suppliers are those who give the greatest value per dollar, after processing, warranty and liability costs are counted." (Chapter 12)

"A defect that is identified during preproduction can be an annoyance. A defect that is identified in the product liability courts can be a disaster." (Chapter 12)

"When your system is in control, don't make changes. Many in-control operations have been thrown out of control when operators made unnecessary changes." (Chapter 13)

"Good briefers know how to emphasize important points without omitting critical parts. They also know how to condense the last 30 minutes of a presentation into a 5-minute summary as soon as the boss starts to fidget." (Chapter 14)

"Four simple illustrations are better than one that is too complex; only the originator may know what it says." (Chapter 14)

"No organization can be confident of its quality unless everyone is contributing." (Chapter 15)

"Defect prevention is the main function of modern quality control, but many companies are not up to date." (Chapter 16)

"When everyone has a stake in the success of the product, they all try to make it a winner." (Chapter 16)

"When witch hunts are no longer the company's favorite recreation, design engineers don't have to worry about being burned at the stake if anything goes wrong. They can quit specifying impossible tolerances." (Chapter 16)

"How much valid information could you get on the marketability of Lincoln limousines from the residents of skid row?" (Chapter 17)

"Those who bypass others create resentment. Bosses who allow bypassing waste their subordinates' time." (Chapter 17)

"Few executives will admit that they don't know as much about quality as anyone else in the plant." (Chapter 17)

"Defect prevention is not a parochial assignment. It's everyone's job." (Chapter 17)

"If you make enough changes, you are sure to stumble on the right one—eventually." (Chapter 18)

"It is hard to correct problems when you don't know they exist." Chapter 18)

"It seldom pays to mount million-dollar campaigns to resolve nickel-and-dime problems." (Chapter 18)

"Although you may never go through detailed calculations like those required in regression analysis, you should be able to interpret the results." (Chapter 18)

"Managers who know nothing about statistics can be as dangerous as a compulsive spender with a dozen corporate credit cards. (Chapter 19)

"It is hard enough to master statistics without having to learn the Greek alphabet." (Chapter 19)

"Computers, with their speedy computations and functional graphics, are great as long as you understand their output." (Chapter 19)

"Specifications should be based on real needs, not impossible dreams." (Chapter 20)

"If you don't know the capability of your process, you are apt to set specifications that are unrealistic." (Chapter 20)

"Specifications that can't be met are like gigantic fortresses made of papier-mache. They don't provide much protection." (Chapter 20).

"Some of the worst quality disasters occur when someone wants to make a minor change like substituting a new raw material or redesigning a functional product. Unfortunately, few people give adequate thought to the question, 'Is the new better than the old?'" (Chapter 21)

"It is much easier to keep from being snowed when you know a little about the language. Statisticians love to snow novices." (Chapter 22)

"With 100 percent inspection, some questionable parts usually slip through; there is always some difference between the vendor's inspection and receiving inspection." (Chapter 23)

"Process control charts are popular tools for monitoring and controlling quality. They reassure you when things are going well and they warn you when things are turning sour." (Chapter 24)

"With most operations, setup is an art." (Chapter 25)

"Pre-Control is a system in which operators control their own processes." (Chapter 25)

"The divide-and-conquer technique is one way of approaching problems systematically." (Chapter 26)

"The easiest way to check a computer program is to give it a problem that has an obvious answer." (Chapter 27)

"Be wary of (regression) analyses that don't make sense. If you make enough analyses, you will eventually find one that appears to be significant, even though it isn't." (Chapter 27)

"Expert statisticians have experimental designs that will confuse almost any manager." (Chapter 28)

"Statisticians seldom use simple words when something more complex will do." (Chapter 28)

"In order to get the most out of an experiment, you should replicate, randomize, and look for interactions between variables." (Chapter 28)

"Without an objective there's no need for an experiment." (Chapter 29)

"The more complex the experiment, the more it will cost and the more it will confuse. Managers seldom accept anything they don't understand." (Chapter 29)

"In the process of running the experiment, unexpected events often occur. Record them." (Chapter 29)

"Retaining an incompetent quality manager is like letting gangrene spread because you don't want to lose a leg." (Chapter 30)

"Quality Managers who know nothing about statistical quality control, vendor quality, specifications, auditing, and inspection can do more damage than a tidal wave hitting Hawaii during a surfers' convention." (Chapter 30)

Index